Qualitative Analysis of Nonsmooth Dynamics

Series Editor
Noël Challamel

Qualitative Analysis of Nonsmooth Dynamics

A Simple Discrete System with Unilateral Contact and Coulomb Friction

Alain Léger
Elaine Pratt

ELSEVIER

First published 2016 in Great Britain and the United States by ISTE Press Ltd and Elsevier Ltd

ISTE Press Ltd
27-37 St George's Road
London SW19 4EU
UK

www.iste.co.uk

Elsevier Ltd
The Boulevard, Langford Lane
Kidlington, Oxford, OX5 1GB
UK

www.elsevier.com

Notices
Knowledge and best practice in this field are constantly changing. As new research and experience broaden our understanding, changes in research methods, professional practices, or medical treatment may become necessary.

Practitioners and researchers must always rely on their own experience and knowledge in evaluating and using any information, methods, compounds, or experiments described herein. In using such information or methods they should be mindful of their own safety and the safety of others, including parties for whom they have a professional responsibility.

To the fullest extent of the law, neither the Publisher nor the authors, contributors, or editors, assume any liability for any injury and/or damage to persons or property as a matter of products liability, negligence or otherwise, or from any use or operation of any methods, products, instructions, or ideas contained in the material herein.

For information on all our publications visit our website at http://store.elsevier.com/

British Library Cataloguing-in-Publication Data
A CIP record for this book is available from the British Library
Library of Congress Cataloging in Publication Data
A catalog record for this book is available from the Library of Congress
ISBN 978-1-78548-094-2

Printed and bound in the UK and US

Contents

Preface

At the beginning of the 20th Century nonlinear dynamics was enriched by many fundamental findings, and from then on was widely studied within the framework of ordinary differential equations. Over the past 60 years, a second-order differential equation with a polynomial nonlinearity and a harmonic forcing, the so-called Duffing equation, was taken as an archetypal model of nonlinear dynamics in discrete systems. Its investigation produced predictive diagrams which are currently used for the occurrence of instability or the transition to chaos.

In the case of partial differential equations the studies are more recent, not as clear and not as complete. It is well known that partial differential equations are classified into different types, namely elliptic, parabolic or hyperbolic, according to the respective orders of the derivative in the different directions, each type leading to very different behaviors. Elliptic partial differential equations are most commonly found to describe the equilibria of solids and structures. In dynamics, parabolic equations may describe the bending deformations of structures but are mainly used in fluids mechanics through the Navier–Stokes equation. Any other motions, whether membrane deformations of structures or dynamics of three-dimensional deformations of solids, are given by hyperbolic equations. Certainly, the nonlinear partial differential equation that has been the most extensively studied is the Navier–Stokes equation, but its behavior remains the topic of many very difficult and technical research studies. In the mechanics of structures, hyperbolic partial differential equations still involve unknowns even concerning basic stability properties.

But the equation of motion changes drastically when the boundary conditions involve conditions of contact or friction, although contact or friction are quite common physical situations. In the case of discrete systems, the equation of motion at first glance still looks like an ordinary differential equation, but a closer look reveals that the equation is far from being classical: the right-hand side involves terms that are not functions, and the equation itself should be understood in the sense of measures. By comparison with classical ordinary differential equations, this feature represents the transition from smooth to nonsmooth dynamics. Very few studies have been performed on this subject. This book book endeavors to investigate the behavior of very simple mechanical systems when the nonsmoothness arises from non-regularized unilateral contact and Coulomb friction.

This book results from a long-term collaboration between the authors. The authors have been working together for the past 15 years, and have also been interested in complementary subjects with respect to the topic of this book; one of the authors has been working for many years on bifurcation and stability problems arising from the mechanics of solids and structures, and the other has been working on nonlinear dynamics for ordinary differential equations.

Over all these years, the authors have been working in the Laboratory of Mechanics and Acoustics of the CNRS in Marseille as a part of the group on Contact Mechanics, whose interests range from theoretical aspects of statics or dynamics to industrial applications, in collaboration with many PhD students.

The authors are highly indebted to Jean-Jacques Moreau; his presence at Montpellier and his regular visits to Marseille were a strong support to the foundations of nonsmooth dynamics in our group. The authors also wish to thank Michel Jean, who built the numerical tools to compute nonsmooth dynamics and to whom the authors owe explicit contributions to the present work. Through an extended analysis of the dynamics of simple systems, this book intends to contribute to the understanding of nonsmooth dynamics. This book is dedicated to the memory of Jean-Jacques Moreau.

Alain LÉGER
Elaine PRATT
Laboratoire de Mécanique
et d'Acoustique, CNRS Marseille
February 2016

Introduction

The Mechanics of Unilateral Systems

A brief glance at the state of the art concerning a large class of problems of mechanics may help us to understand the objectives of the present book. From a very wide and very abstract point of view, the statement of a problem of solid mechanics requires a specific combination of concepts coming from physics. For example, does the physics require that the problem be modeled by a discrete system or by a continuous medium? Does it imply that the boundary conditions are unilateral or bilateral? Do these conditions involve contact and friction? Is the problem static or dynamical? The complexity of these alternatives may rapidly increase by considering subclasses also arising from physics, for example does the system undergo large deformations or small deformations, involve distance interactions or contact interactions.

An initial outline of the present book would be that it considers a specific combination of the following alternatives: the model is discrete or continuous, the problem is static or dynamical, involves unilateral contact, with or without friction, and in each case, focus will be concentrated on the question of utmost physical interest, namely, the question of whether there exists a solution to the corresponding mathematical model, and moreover does there exist a single solution.

1) In the case of discrete systems:

i) The static problem of a discrete system with or without friction is now well understood, but it is important to bear in mind that although an equilibrium problem may look simple, the theoretical results have been established quite recently and even now justify an extended analysis.

ii) The dynamical problem for a discrete system has been completely solved only recently in the case without friction. After the first fundamental existence result given by Michelle Schatzman in 1978, the well-posedness was obtained for a general problem in $\mathbb{R}^n$ without friction under strong restrictions on the regularity of the loading, by Patrick Ballard in 2000. The case with friction is precisely the subject of this book. The dynamical response is relatively well understood only for simple models, but remains intricate for more general problems, although well-posedness for large-size systems was obtained in 2014 under conditions on the loading similar to those concerning the case without friction.

2) In the case of linearly elastic continuous media the situation is totally different, and way beyond the scope of this book:

i) The formulation of the equilibrium problem of a linearly elastic solid at small strains submitted to unilateral contact conditions without friction is due to Antonio Signorini [SIG 59] in the 1950's, and Gaetano Fichera [FIC 63] proved in the 1960's existence and uniqueness for this equilibrium problem. Later on in the 1960s, these results were generalized within the new framework of variational inequalities. Uniqueness is lost in general if the framework of linear elasticity at small strains is removed and replaced either by nonlinear elasticity, by large strains, or by any other nonlinear behavior. However, unilateral conditions at the boundary are not in general cause of this loss of uniqueness.

ii) No similar result exists if the conditions at the boundary involve Coulomb friction even in the static case where the problem remains open and certainly difficult.

iii) Almost all the problems related to the dynamics of a continuous medium in the presence of unilateral contact are open for the most part. There are no results at all in the presence of Coulomb friction. In the case of contact without friction, only two results can be viewed as first steps toward solving the problem. The first one is the work by G. Lebeau and M. Schatzman who obtained existence and uniqueness of an energy preserving solution to the wave equation, but their result was strictly restricted to a half space with unilateral contact at the boundary. The second one was obtained by J. U. Kim, as opposed to the result by G. Lebeau and M. Schatzman, it dealt with a bounded domain of any smooth enough shape. However, there was a strong restriction: it did not apply to elasticity but only to the harmonic operator, and moreover established existence but not uniqueness. Of course, an existence result can be seen as an important first step and is often the most difficult step, but here it was considered in the mechanical community as being of little physical interest as it dealt only with the Laplacian.

3) Are the investigation of the behavior and the computation of trajectories of solutions to continuous systems involving unilateral contact and Coulomb friction possible at short term?

i) *Regularizing contact or friction laws*: The contact and friction laws will be discussed in Chapter 2, but the essential point is that these laws are represented by multivalued mapping and not by functions. Regularizing these laws means changing these multivalued mappings into functions, but no convergence of the regularized solution toward the solution of the corresponding mechanical problem has been established and indeed no estimation of the error is available because there are as yet no results giving existence and uniqueness of the solution under unilateral contact and friction. Therefore, regularization does not seem to be an interesting path to follow.

ii) *Discretizing continuous bodies*: Another idea, currently used in commercial softwares, comes as a trivial consequence of points (1) and (2) above: in the case of discrete systems, exploring the set of equilibrium states or the dynamics has either already been performed or is possible, while almost all the problems whether static or dynamical are open in the case of continuous media. Then, the natural idea was to discretize the continuous media at hand, by a finite element method for example, and to perform the calculations using the corresponding discrete problem. The use of a finite element method or of any other discretization of a continuous medium requires the convergence of the solution, which means that the solution should tend to the solution for the continuous body when discretization step size, for example mesh size in the case of finite elements, tends to zero. Unfortunately, such a convergence result does not exist when the continuous body is submitted to unilateral contact and to Coulomb friction and, as in the case of regularization, establishing such a convergence result would lead to the same difficulties as those encountered when tackling the continuous problem itself.

iii) *Restricting attention to discrete systems*: Discrete systems involve one more equation than continuous bodies, essentially because the discrete model removes the possibility of waves propagating inside the body after an impact time, but discrete models, nevertheless, exhibit a very large range of behaviors that are far from being well known. Even if studying discrete models is a strong restriction with regard to the analysis of the mechanics of unilateral systems in general, they will certainly bring important insight on nonsmoothness, whether in statics or dynamics. The authors have decided in the present book to restrict their attention to discrete systems.

The book is organized as follows:

– Chapter 2 aims at describing the models. The description first focuses on the constitutive laws. The meaning of the unilateral contact and friction laws will be given in detail. Then, the finite degree of freedom mechanical devices that will be studied in the following chapters will be presented, either at small or at large deformations.

– Chapter 3 is concerned with theoretical results and numerical tools. After setting the mathematical foundations of the dynamical problem of a discrete system submitted, in addition to external forces, to unilateral contact and Coulomb friction with some rigid support, the current state of theoretical results for this problem is given. In particular, the necessary conditions insuring uniqueness are stated and counterexamples to uniqueness are given. Then, a number of numerical methods designed to harness the specificity of the dynamics are given. Specificity means that unilateral contact implies, for instance, that a trajectory may involve impacts and jumps; Coulomb friction may signify that a given particle is sliding on the support during a time interval, while it is stuck by friction, as if it were clamped, during another time interval. These particularities appear naturally from the functional framework and must be taken correctly into account in the computational methods.

– The investigation of the behavior of the solution is given in Chapter 4. In the case where the restoring force is linear, referred to as the linear case, it is shown that the set of equilibrium states depends only on two parameters. One of these parameters couples the stiffness parameters with the external force, the other couples the stiffness parameters with the friction coefficient. Depending on whether they are positive, negative or equal to zero, there exist (or not) equilibrium states out of contact, and when there exist equilibrium states in contact, there are infinitely many except in one particular case. In the case of large deformations, referred to as the nonlinear case, the investigation is much more complex and cannot be represented by a simple two-parameter table, but it is nevertheless extensively described. Again, the results of this investigation would be drastically changed by any regularization.

– Chapter 5 discusses the question of whether the equilibrium states that have been obtained in the previous chapter are stable or not. The question arises naturally because the stability theory of classical dynamical systems suggests that often when several equilibrium states are obtained, one is stable and the others are not. Sometimes it is not so simple, but here the generic situation consists of infinitely many equilibrium states, so that it is important to have information about their stability. A first answer is given by the study of the trajectories: whether all the trajectories starting from initial data in a neighborhood of an equilibrium remain close to the equilibrium for any time or whether one initial data in such a neighborhood lead to a trajectory that diverges from the equilibrium. All the equilibria in the linear case have been explored in this way. But it was observed that, due to Coulomb friction, this classical stability analysis may not be suitable to the physical requirements. A new definition of stability, specially suited to Coulomb friction, is then proposed and a stability conjecture is given, which is backed up by direct calculations of the final states reached by the trajectories of simple systems.

– The analyses provided in Chapters 6 and 7 are exhaustive and constitute the most original part of the book. Chapter 6 deals with the linear case when the system is submitted to a periodic loading. It is obtained that the plane $\{\text{period T}, \text{amplitude } \varepsilon\}$ of the excitation is essentially divided into three zones: the first one is a horizontal strip $T \in]0, +\infty[\times \varepsilon \in]0, \varepsilon_0[$ in which there exist infinitely many equilibrium states and no periodic solutions. The second range has a lower boundary at the upper boundary of the first one and a rather complicated upper boundary. Here, there exist periodic solutions but no equilibrium states. Above the complicated boundary there no longer exist sliding solutions, which means that all the trajectories involve jumps and impacts.

Due to the fact that the set of equilibria of the nonlinear case is much more intricate than in the linear case, a complete partition of the {period T, amplitude ε} plane is not possible. The problem tackled in Chapter 7 is of a more qualitative nature and focuses on the effect of the coupling between smooth and nonsmooth nonlinearities on the dynamics of low-dimensional dynamical systems. The nonsmooth nonlinearities result from unilateral contact and Coulomb friction while the smooth ones deal with large deformations. The whole range of periods is explored and a qualitative result is that a complicated dynamics leading to chaos appears.

– Partly as concluding remarks, as an opening toward remaining problems and challenges, Chapter 8 updates the state of the art that has been given in Chapter 1. Among challenging problems that arise from the analysis presented in this book, emphasis is put on the following points:

- the necessity of a general theory of stability for unilateral systems;

- the improvement of the well-posedness criterion of the dynamics of discrete systems;

- the mathematical problems of the transition from discrete to continuous systems that have been raised in Chapter 1.

The Model

This chapter contains the different mechanical models to which the analysis presented in this book is applied. Two types of nonlinearities are considered here. Some arise from the constitutive laws others result from the geometry of the models.

The contact and friction conditions are referred to as nonsmooth nonlinearities, by which it is meant that they involve multivalued mappings. No regularization of these nonsmooth nonlinearities are carried out.

The nonlinearities resulting from the geometry are smooth functions. Only discrete systems, in fact essentially very simple mass–spring models with linearly elastic springs, are considered so that there are no nonlinearities arising from the material properties.

Therefore, the mechanical problems that will be tackled deal with the coupling between nonsmooth nonlinearities and smooth functions (which can even be linear in some cases). Different models are presented, ranging from very simple to large size systems and from small to large deformations.

A formal expression of the equations governing the dynamical problem concludes the chapter. The explicit expression of these equations will be given in the following chapters.

1.1. About contact and friction conditions: nonsmooth nonlinearities

Using a model as simple as possible, the contact and friction conditions are classically divided into two parts, one concerning the tangential components, and the other concerning the normal components, where tangential and normal refer in the space $\mathbb{R}^2$ to a local frame $(\overrightarrow{t}, \overrightarrow{n})$. The tangent vector $\overrightarrow{t}$ is attached to the line the particle is impacting or sliding on; the normal vector $\overrightarrow{n}$ is directed so as to have the admissible space for the motion in the negative half plane. Let U with components (u_t, u_n) be the displacement of the particle and R with components (R_t, R_n) be the reaction of the line, which is identified with an obstacle. These conditions read as follows:

– For the normal components:

$$u_n \leq 0, \quad R_n \leq 0, \quad u_n R_n = 0. \tag{1.1}$$

This set of relations is classically referred to as Signorini conditions or unilateral contact conditions. It means that the particle must be on the right side of the obstacle, that the reaction of the obstacle can only push the particle away, and that a non-zero reaction is possible only when the particle is in contact with the obstacle.

Concerning the tangential components, the model chosen all along this book is the so-called Coulomb friction law. It involves a positive real coefficient, the so-called friction coefficient, in the following way:

$$\begin{cases} R_n = 0 \implies \dot{u}_t \in \mathbb{R}, \\ \mu R_n \leq R_t \leq -\mu R_n, \\ \text{with } \begin{cases} |R_t| < -\mu R_n \implies \dot{u}_t = 0, \\ |R_t| = -\mu R_n \implies \exists \lambda \geq 0 \text{ s.t. } \dot{u}_t = -\lambda R_t. \end{cases} \end{cases} \tag{1.2}$$

These conditions mean that no sliding is possible while the tangential component of the reaction belongs strictly to an interval the ends of which are proportional to the normal component of the reaction, and that the particle can be set into sliding motion only if the tangential component of the reaction reaches one end of this interval.

Signorini conditions and Coulomb's friction conditions are represented by their graphs in Figure 1.1.

In a mechanical model, the reaction of the obstacle is a force, which will consequently be introduced in the same way as the given external forces or the internal stresses in the right-hand side of the equation of the dynamics. The vertical parts of the graphs in Figure 1.1 make it quite clear that these contact and friction conditions cannot be expressed by a function connecting the reaction to the position or to the velocity as a constitutive law would do in general. This particularity will make the dynamical system very different from usual ordinary differential equations. Many studies and corresponding softwares change the graphs of these conditions, respectively, in the $\{u_n, R_n\}$ plane (for contact) and in the $\{\dot{u}_t, R_t\}$ plane (for friction) into graphs of ordinary functions by some smoothing procedure. It is clear that these regularized graphs lead to a huge simplification of the mathematics, especially in dynamics, but also for equilibrium problems. However, the outcome obviously depends on the smoothing that has been adopted, and, whereas the non-regularized relations [1.1] and [1.2] have a true physical meaning (non-penetrability, sliding only for large enough force, etc.), the most common regularizations of the contact or friction laws are clearly not physical. Indeed, two examples of regularizations of Coulomb's law among many others are represented in Figure 1.2. With such a regularization, an infant would manage to initiate a sliding motion in a one ton block. The main asset of these regularizations is that the framework of multivalued mappings is replaced by that of smooth functions so that the models are governed only by ordinary differential equations. But the important drawback is that specific behaviors are brought to light through the non-regularized laws that would not appear if a regularization had been operated.

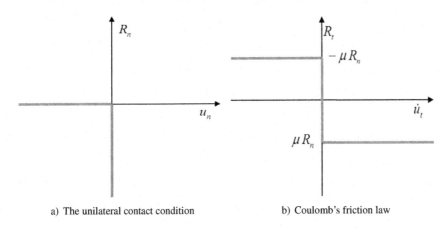

a) The unilateral contact condition   b) Coulomb's friction law

Figure 1.1. *The graph of the contact and friction conditions*

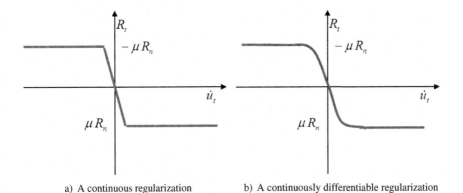

a) A continuous regularization   b) A continuously differentiable regularization

Figure 1.2. *Two regularizations of Coulomb's friction law*

It is interesting to merge the two graphs in Figure 1.1 into a single one in the $\{R_t, R_n\}$ plane. Admissible values of the reaction cannot belong to the half plane of positive R_n as shown in Figure 1.1(a), nor can they belong to the two parts of the half plane of negative R_n corresponding to $|R_n| > \frac{1}{\mu}|R_t|$, as shown in Figure 1.1(b), so that an admissible value of the reaction should belong to the part of the plane, which is unshaded as shown in Figure 1.3 and which will from now onwards be referred to as the Coulomb cone. This remark actually holds as a constitutive law: as long as the reaction is strictly inside this zone, the particle is strictly stuck by friction; a sliding motion is possible only when the reaction is on the boundaries of the cone, that is on the two half-lines $R_n = \pm\frac{1}{\mu}R_t$, and loosing contact implies passing through the vertex of the cone.

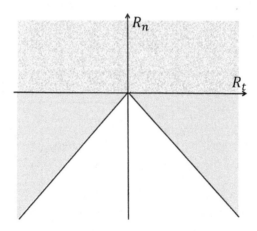

Figure 1.3. *The contact constitutive law represented in the space of the reactions*

1.2. A class of discrete systems

1.2.1. *The basic mass–spring systems*

The basic model is built in the following way: a mass m is moving in the $\mathbb{R}^2$ plane where it is connected to a rigid frame by linearly elastic springs, but in addition to the effects of the spring, the mass is constrained to remain in a half-plane due to the presence of an obstacle represented by the horizontal line. Contact and friction with this obstacle is assumed to hold in a strictly unilateral way as given by equations [1.1] and [1.2]. This kind of system was first introduced in order to investigate the loss of existence or uniqueness of quasi-static evolution in presence of Coulomb friction [KLA 90], with the long-term aim of understanding the behavior of finite element discretizations of continuous bodies. Many such small-size systems like those shown in Figure 1.4 can be built and are found in the literature. The following system, shown in Figure 1.5, was introduced in the 1980s for computational purposes and for the analysis of uniqueness of the solution [ALA 86]. It will also be used here in a discussion about stability.

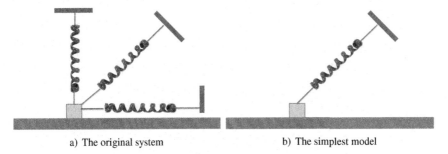

a) The original system   b) The simplest model

Figure 1.4. *Simple models*

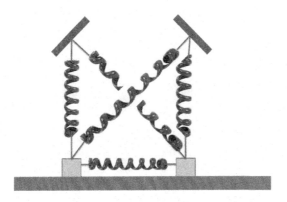

Figure 1.5. *A slightly more complicated system*

1.2.2. *A special case without coupling*

Another kind of mechanical device, highly classical in the physics of solids, will also be used. The main particularity of this device is that the coupling between normal and tangential components is removed due to the fact that the normal component of the force is given, the weight for example, which means that the normal component of the reaction R_n in equation [1.2] is replaced by a given constant, or, in other words, the length or the interval $[\mu R_n, -\mu R_n]$ is given and no longer unknown. Such a model is shown in Figure 1.6 and contains any number of particles.

Figure 1.6. *A chain of mass with Coulomb friction*

1.2.3. *Larger size systems*

The following two kinds of much larger size systems have been investigated:

– the first one is the finite dimensional model obtained when discretizing any continuous body into finite elements as shown in Figure 1.7. The previous elementary systems as shown in Figure 1.4 can be seen as the simplest model of a two-dimensional (2D) finite element and have been introduced to help explain the behavior of any continuous body in the presence of Coulomb friction;

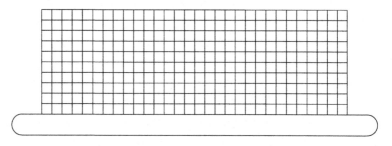

Figure 1.7. *A finite element discretization of a deformable solid resting on a support*

– the second large system has the particularity that the stiffness matrix reduces to zero. It is usually of very large in size, grains of sand in a box or in a pile for example, and is such that unilateral contact and friction conditions hold everywhere between each grain. Very simple geometries can be used (Figure 1.8(a)), but usually grains of any polygonal shapes are required (Figure 1.8(b)).

1.3. Study of the restoring force

Except when explicitly specified, attention is now restricted to the simple mass spring system, which involves unilateral contact and Coulomb friction, the full coupling between normal and tangential components and is studied either at small or at large deformations.

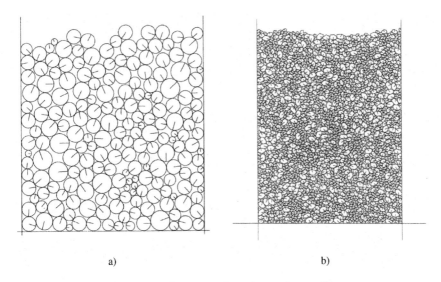

Figure 1.8. *Two examples of granular media*

1.3.1. *The geometrically linear case*

The mechanical device is shown in Figure 1.9, which, in addition to Figure 1.4(b), includes some notations. The spring is assumed to be linearly elastic of stiffness k, of length l and undergoing simple deformations of extension or compression where the changes in the spring's direction are neglected. Let $U = (u_t, u_n)$ be the displacement vector of the mass expressed in the local frame $(\overrightarrow{t}, \overrightarrow{n})$ and let $\mathcal{F}(U)$ be the restoring force associated with an extension U of the spring. The strain energy of the system is given by

$$\mathcal{W}(U) = \frac{1}{2}kU^2 = \frac{1}{2}k[u_t \sin\varphi + u_n \cos\varphi]^2. \qquad [1.3]$$

As the direction of the spring does not change the angle, φ is constant so that this quadratic energy can be rewritten as

$$\mathcal{W}(U) = \frac{1}{2}\left(u_t, u_n\right) K \begin{pmatrix} u_t \\ u_n \end{pmatrix}, \qquad [1.4]$$

where K stands for the stiffness matrix now written as

$$K = \begin{pmatrix} K_t & W \\ W & K_n \end{pmatrix}.$$

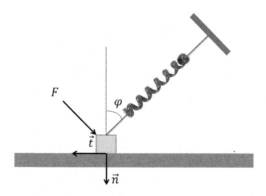

Figure 1.9. *The geometrically linear simple mass–spring system*

The coefficients of matrix K are easily identified from equation [1.3], which leads to

$$K = k \begin{pmatrix} \sin \varphi^2 & 2 \sin \varphi \cos \varphi \\ 2 \sin \varphi \cos \varphi & \cos \varphi^2 \end{pmatrix},$$

so that the restoring force is

$$\mathcal{F}(U) = -KU = \begin{cases} -K_t u_t - W u_n &= -k \sin \varphi^2 u_t - 2k \sin \varphi \cos \varphi u_n \\ -W u_t - K_n u_n &= -2k \sin \varphi \cos \varphi u_t - k \cos \varphi^2 u_t. \end{cases} \quad [1.5]$$

This expression would be practically identical for the system in Figure 1.4(a), in particular the coupling by the non-diagonal term $W = 2 \sin \varphi \cos \varphi$ would remain; only the diagonal terms would be changed as follows:

$$K_t = k(\sin \varphi^2 + 1), \quad K_n = k(\cos \varphi^2 + 1).$$

Any other simple bi-dimensional system would be modeled in the same way where the stiffness matrix K would remain positive definite.

1.3.2. *The geometrical nonlinearity in 2D: adding smooth nonlinearities*

When the change in the direction of the spring cannot be neglected, situation referred to as large deformations, the restoring force $\mathcal{F}(U)$ associated with an extension U will be written as $\mathcal{N}(U)$. Since the displacement U is now assumed to

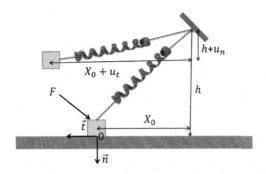

Figure 1.10. *Calculation of the nonlinear strain*

have any direction, the simple system shown in Figure 1.4(b) is now represented in Figure 1.10.

The components of the restoring force are written as:

$$\mathcal{N}(\mathbf{U}) = \left\{ \begin{array}{l} \mathcal{N}_t(u_t, u_n), \\ \mathcal{N}_n(u_t, u_n). \end{array} \right. \tag{1.6}$$

Assume that the mass is initially in contact with the obstacle (which is in no way restrictive). Using the notations given in Figure 1.10, $\sqrt{X_0^2 + h^2}$ is the natural length of the spring. A displacement of components (u_t, u_n) changes the length of the spring from $\sqrt{X_0^2 + h^2}$ to $\sqrt{(X_0 + u_t)^2 + (h + u_n)^2}$. The spring still being of stiffness k, the strain energy of the system is:

$$\widetilde{\mathcal{W}}(\mathbf{U}) = \frac{1}{2}k\left[\sqrt{(X_0 + u_t)^2 + (h + u_n)^2} - \sqrt{X_0^2 + h^2}\right]^2, \tag{1.7}$$

from which the components of the restoring force can be obtained:

$$\left\{ \begin{array}{l} \mathcal{N}_t(u_t, u_n) = -k(X_0 + u_t) \\ \qquad + \dfrac{k(X_0 + u_t)\sqrt{X_0^2 + h^2}}{\sqrt{(X_0 + u_t)^2 + (h + u_n)^2}}, \\[3mm] \mathcal{N}_n(u_t, u_n) = -k(h + u_n) \\ \qquad + \dfrac{k(h + u_n)\sqrt{X_0^2 + h^2}}{\sqrt{(X_0 + u_t)^2 + (h + u_n)^2}}. \end{array} \right. \tag{1.8}$$

1.4. The dynamical problem

The dynamical problem can now be written. Assume the mass is submitted to an external force F of components (F_t, F_n), then the trajectories will be solutions of a system of the form:

$$\left\{ \begin{array}{l} m\ddot{U} = \mathcal{F}(U) + F + R, \\[4pt] \text{Initial data,} \\[4pt] \text{Unilateral contact conditions,} \\[4pt] \text{Coulomb's friction conditions,} \\[4pt] \text{Impact law.} \end{array} \right. \qquad [1.9]$$

The meaning and the explicit form of the last line of system [1.9] will be discussed later on. System [1.9] can be written more explicitly as

$$\left\{ \begin{array}{l} \text{i)} \left\{ \begin{array}{l} m\ddot{u}_t = \mathcal{F}_t(u_t, u_n) + F_t + R_t, \\[4pt] m\ddot{u}_n = \mathcal{F}_n(u_t, u_n) + F_n + R_n, \end{array} \right. \quad t > 0 \\[20pt] \text{ii)}\ u(0) = u_0,\ \dot{u}(0) = v_0, \\[8pt] \text{iii)}\ u_n \leq 0,\ R_n \leq 0,\ u_n R_n = 0, \\[8pt] \text{iv)} \left\{ \begin{array}{l} R_n = 0 \implies \dot{u}_t \in \mathbb{R}, \\[4pt] \mu R_n \leq R_t \leq -\mu R_n, \\[4pt] \text{with} \left\{ \begin{array}{l} |R_t| < -\mu R_n \implies \dot{u}_t = 0, \\[4pt] |R_t| = -\mu R_n \implies \exists \lambda \geq 0 \text{ s.t. } \dot{u}_t = -\lambda R_t, \end{array} \right. \end{array} \right. \\[28pt] \text{v) Impact law.} \end{array} \right. \qquad [1.10]$$

In the case of small deformations, the restoring force is $\mathcal{F}(U) = -KU$, so that equation [1.10(i)] becomes:

$$\left\{ \begin{array}{l} m\ddot{u}_t + K_t u_t + W u_n = F_t + R_t, \\[4pt] m\ddot{u}_n + W u_t + K_n u_n = F_n + R_n. \end{array} \right. \quad t > 0$$

In the case of large deformations, the restoring force is $\mathcal{F}(U) = \mathcal{N}(U)$ so that equation [1.10(i)] becomes:

$$\begin{cases} m\ddot{u}_t = \mathcal{N}_t(u_t, u_n) + F_t + R_t, \\ \\ m\ddot{u}_n = \mathcal{N}_n(u_t, u_n) + F_n + R_n. \end{cases} \qquad t > 0$$

In both cases, the unknowns consist of the four components u_t, u_n, R_t and R_n. Theoretical and numerical tools for the analysis of system [1.10] are presented in the following chapter.

Mathematical Formulation

2.1. The complete mathematical statement

The term nonsmooth nonlinearity has already been introduced in Chapter 2. However, it is important to give a more accurate characterization of nonsmoothness by highlighting two points:

– the first point is discussed in Chapter 2. The graphs shown in Figure 1.1 were introduced as constitutive laws, which usually means that the reaction of the obstacle is given by a function of the displacement, so the displacement is the single unknown of the equilibrium or dynamical problems, as it is the case in elasticity, whether linear or not. But specifically the graphs in Figure 1.1 are not those of functions. This is not exceptional, it occurs also, for example, in plasticity, but the effects of this nonsmoothness are important whether in statics or in dynamics, in particular the equilibrium problem may have infinitely many solutions;

– the second point is related to dynamics and is qualitatively illustrated in Figure 2.1. Assume that a single particle is moving in the plane when submitted to a force $F(t)$, and that a part of this plane is forbidden to the particle. The curve representing the boundary of this part of the plane can consequently be seen as an obstacle: the particle must remain on the same side of this obstacle. Then, either the particle is not in contact with the obstacle and its velocity is a continuous function of time whatever the smooth function $F(t)$, or the particle enters into contact with the obstacle, and there is a jump in the velocity at the time of impact.

This implies that whatever the function $F(t)$, the acceleration is the derivative of a function that is not continuous. According to these remarks about the nonsmoothness and its implications, it is now possible to give a rigorous mathematical statement of the dynamics. It reads as:

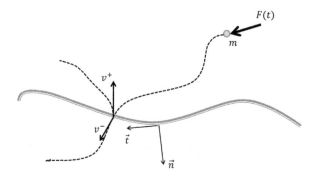

Figure 2.1. *Example of a trajectory containing an impact time*

Find $U \in \mathrm{MMA}([0,T];\mathbb{R}^n)$ and $R \in \mathcal{M}([0,T];\mathbb{R}^n), n = 2,3,$ such that

$$
\begin{cases}
\text{i) } m\ddot{U} = \mathcal{F}(U) + F + R, \quad t > 0, \\[2mm]
\text{ii) } U(0) = U_0,\ \dot{U}^+(0) = V_0, \\[2mm]
\text{iii) } u_n \leq 0,\ R_n \leq 0,\ u_n R_n = 0, \\[2mm]
\text{iv) } \begin{cases} \forall \varphi \in \mathbb{C}^0([0,T];\mathbb{R}^{n-1}), \\[2mm] \displaystyle\int_{[0,T]} R_t(\varphi - \dot{u}_t^+) - \mu R_n(|\varphi| - |\dot{u}_t^+|) \geq 0 \end{cases} \\[4mm]
\text{v) } u_n(t) = 0 \implies \exists e \in [0,1] \text{ such that } \ddot{u}_n^+(t) = -e\ddot{u}_n^-(t).
\end{cases}
\qquad [2.1]
$$

The generic term $\mathcal{F}(U)$ represents either $\mathcal{N}(U)$ in the nonlinear case or $-KU$ in the linear case, as introduced in Chapter 2. Due to physical requirements, the displacement must be a continuous function. The notation $\mathrm{MMA}([0,T];\mathbb{R}^n)$ for the functional framework is due to Michelle Schatzman. MMA stands for motion with measure acceleration, so $\mathrm{MMA}([0,T];\mathbb{R}^n)$ stands for the set of continuous functions the second derivative of which is a measure, defined on $[0,T]$ with values in $\mathbb{R}^n$.

A few additional comments are necessary about system [2.1]:

1) $\dot{U} \notin \mathbb{C}^0([0,T])$ implies that $\ddot{U} \in \mathcal{D}'([0,T])$ so equation [2.1(i)] implies that the reaction R also belongs to $\mathcal{D}'([0,T])$. But the reaction must satisfy a sign condition given by [2.1(iii)], which implies that R is a measure, so the equation of the dynamics should be taken in the sense of measures (see [SCH 78] and [MOR 88]).

2) The next comment is a consequence of the previous comment: since the equations of the dynamics should be understood in the sense of measures, they can be integrated over an interval, as a basic property of a real-valued measure. This feature is essential in the proof of an existence result (see [MON 93]) and of a numerical method for the calculation of the trajectory (see [JEA 99]). Moreover, since the acceleration is a measure, its primitive, which is the velocity, belongs to BV, the set of functions of bounded variation, of which a basic property is that it has right and a left limits at any time, and this property is exactly what is needed: the right velocity at the origin $\dot{u}_n^+(0)$ is well defined and can be stated as equal to some given value v_0 (see [MOR 88]), while the formal $\dot{U}(0) = V_0$ would have no meaning; furthermore, the situation shown in Figure 2.1 does not require the velocity to exist at any time; it only requires that its left and right limits exist at any time in order to give a meaning to the impact law [2.1(v)].

3) Equation [2.1(iv)] is a reformulation of friction conditions, which is necessary because the reaction is a measure and the velocity a function of bounded variation. Indeed, conditions [1.2] are formal but [2.1(iv)] is perfectly defined due to the following comments:

- formulation [1.2] is recovered by integrating by parts [2.1(iv)] and conversely [1.2] leads to [2.1(iv)];

- since the set of measures is the dual of the set of continuous functions, integrals over $[0, T]$ of terms such as $\varphi.R$ are duality products;

- a function of bounded variation can be integrated with respect to any measure so integrals over $[0, T]$ of terms such as $R.\dot{u}^+$ are also well defined since R is a measure.

4) The impact condition [2.1(v)] is necessary to ensure that problem [2.1] has a solution after an impact time as would be the case for any discrete system. This condition would not have to be introduced in the dynamics of continuous bodies due to the fact that an impact on the boundary leads to the propagation of waves inside the body. It is, however, necessary for the idealization of a discrete system made up of any collection of single points or rigid bodies. It involves a so-called restitution coefficient, denoted by e in system [2.1], the value of which is determined by physics, and is well defined because the right and left limits of the velocity exist. The case $e = 1$ is called perfectly elastic, while the case $e = 0$ defines a perfectly inelastic impact.

5) The friction conditions induce a dissipative behavior. All the qualitative properties given here would not be changed if an additional and more classical damping term was added.

Attention must also be drawn to the fact that problem [2.1] has four unknowns that are u_t, u_n, R_t and R_n, while it involves two equations and a set of conditions given mostly by inequalities. This makes it obvious that the dynamical system is qualitatively very different from ordinary differential equations. Due to the above comments, it is then natural to address the question of the well-posedness of the dynamics, so the remaining part of section 2.1 is organized as follows:

– the theoretical results ensuring the existence of a solution to problem [2.1] are first recalled. The corresponding section is reduced to recalling the results since these existence theorems together with their proofs have already been published;

– it has then been proved that within the framework where existence has been established, the Cauchy problem [2.1] is ill-posed and would remain ill-posed even within a much restricted framework. Three counterexamples to uniqueness are presented here. Although the first of these counterexamples has been presented by several authors, the three counterexamples concern problems that differ sufficiently for one not to be deduced straightforwardly from the other. Moreover, one of them has never been published;

– the last section details the conditions for the Cauchy problem to be well posed. It has been established that in a still more restricted functional framework, uniqueness is recovered. This result has been proved by different methods, so its presentation in this chapter consists of its exact statement and of a brief survey of one of the proofs.

2.1.1. *Existence results*

The first proof that the non-regularized impact problem has a solution was obtained by Schatzman [SCH 78] in 1978, in the frictionless case and for a single mass moving in the presence of an obstacle represented by a rigid wall. The external force was assumed to depend only on time and the impact to be perfectly elastic. The main tool for this proof was a penalty method in which the limit problem as the penetration of the impacting mass into the wall tends to zero still has a solution that belongs to $\mathbb{MMA}([0,T];\mathbb{R})$. Later on Schatzman proved the existence of a solution in the case where the external force is not only function of time but also of the displacement and the velocity [SCH 98].

In the case involving Coulomb friction, the existence of a solution was proved by Monteiro Marquès [MON 93] who studied a two degree of freedom model very close to the one shown in Figure 1.4. His proof was based upon the convergence of a time discretization. After building approximate problems associated with sequences of forces, accelerations, velocities and displacements, he was able to prove, due to compactness arguments, that the corresponding sequence of solutions has a limit given by the convergence of the displacements in $\mathbb{C}^0$, the convergence of the velocities in $\mathbb{BV}$, of the accelerations in measure, and the fact that this limit satisfies all the conditions of the initial time continuous problem. First given for an essentially bounded external force, this result holds in fact as soon as the external force is integrable. The time discretization used in this proof is exactly the one described in the section dealing with computational techniques, so its presentation can be found at the end of this chapter when the so-called time-stepping method is given.

For larger systems, Ballard [BAL 00] first extended the result of the frictionless case to any finite dimensional system involving unilateral contact. The idea which was at the root of the proof was to compare the problem with the corresponding bilateral problem, which involves only an ordinary differential equation, and then to establish that a locally analytic solution exists in a right neighborhood of the origin of time if the external force is analytic. Then, he showed that the same tools apply for the two degree of freedom problem in Figure 1.4 [BAL 05], which he extended to any system in $\mathbb{R}^n$ with unilateral contact, Coulomb friction and a forcing depending analytically on time and Lipschitz continuous in displacements [CHA 14].

2.1.2. *Counterexamples to uniqueness*

2.1.2.1. *The case with perfectly elastic bounces*

This section recalls in detail the first counterexample to uniqueness that was given by Schatzman in 1978. The idea was to show through explicit analytical calculations that a $C^\infty([0,T])$ external force can be built such that problem [2.1] has two different solutions. The problem is described in Figure 2.2. From a mechanical point of view, it is as simple as possible: a single particle can move in a single direction perpendicular to a rigid wall; when it is not in contact with the wall, the particle is submitted only to a given force; when it is in contact with the wall, there is a reaction according to equation [2.1(iii)]; the impacts on the wall are assumed to be perfectly elastic, which means that $e = 1$ in equation [2.1(v)], so the dynamics is governed by

$$
\left\{
\begin{array}{l}
\text{Find } U \in \mathbb{MMA}([0,T];\mathbb{R}) \text{ and } R \in \mathcal{M}([0,T];\mathbb{R}), \text{ such that} \\[2mm]
\text{i) } m\ddot{U} = F + R, \quad t > 0, \\[2mm]
\text{ii) } U(0) = 0, \ \dot{U}^+(0) = 0, \\[2mm]
\text{iii) } U \le 0, \ R \le 0, \ UR = 0, \\[2mm]
\text{iv) } U(t) = 0 \Longrightarrow \dot{U}^+(t) = -\dot{U}^-(t).
\end{array}
\right.
\qquad [2.2]
$$

In order to be sure that a trivial solution exists, the particle is at rest against the wall before time $t = 0$ and the loading is positive, or in other words the force always pushes the particle against the wall, which actually means that $U \equiv 0$ with a reaction R opposite to the loading will always be a solution. So a smooth positive loading that would induce a non-zero solution is looked for. The tool to do this is the so-called Massin function defined as

$$\rho : \mathbb{R} \longrightarrow \mathbb{R} \qquad t \mapsto \begin{cases} 0, & \text{if } t \in \,]-\infty, 0] \cup [1, +\infty[, \\[2ex] \left[\displaystyle\int_0^1 \dfrac{1}{e^{t\,(t-1)}}\,dt \right]^{-1} \dfrac{1}{e^{t\,(t-1)}} & \text{if } t \in \,]0, 1[, \end{cases} \qquad [2.3]$$

which satisfies the following properties:

$$\begin{aligned} & \rho \in \mathcal{C}^\infty(\mathbb{R}; \mathbb{R}^+), \quad \operatorname{supp} \rho \subset [0, 1], \\[1ex] & \forall n \in \mathbb{N}, \ \frac{d^n}{dt^n}\rho(0) = \frac{d^n}{dt^n}\rho(1) = 0, \\[1ex] & \int_0^1 \rho(s)\,ds = 1, \quad \int_0^1 s\rho(s)\,ds = \frac{1}{2}, \end{aligned} \qquad [2.4]$$

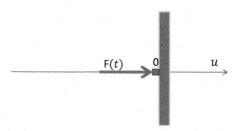

Figure 2.2. *The particle against the wall*

It is consequently a $\mathbb{C}^\infty([0,1])$ bell-shaped function on a bounded interval with $\mathbb{C}^\infty$ matches with zero on both sides of this interval, as shown in Figure 2.3. Now the idea is to build a sequence $\{a_n\}_{n\in\mathbb{N}}$ strictly decreasing to zero and to choose time $\hat{T}$ equal to a_0 so the interval $[0, \hat{T}[$ is divided into sub-intervals $[a_{n+1}, a_n[$. Then, the loading F is defined by

$$F(0) = 0,$$

$$F(t) = \begin{cases} 0, & \text{if } t \in [a_{n+1}, a_{n+1} + d_n[, \\[2ex] \dfrac{F_n}{2}\rho\!\left(\dfrac{t - a_{n+1} - d_n}{a_n - a_{n+1} - d_n}\right) & \text{if } t \in [a_{n+1} + d_n, a_n[, \end{cases} \qquad [2.5]$$

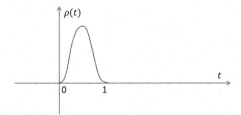

Figure 2.3. *Massin function*

The function $\rho(.)$ has been defined by equation [2.3], so by choosing $F_n = \dfrac{1}{n!}$ the function F is $\mathbb{C}^\infty(]0, \hat{T}[)$. In fact, it can be checked by direct calculations that all the derivatives of F are equal to zero at $t = 0$ because of this particular choice of F_n. The loading F is shown in Figure 2.4.

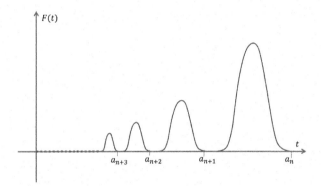

Figure 2.4. *The loading built with Massin functions*

Then, the following choices are made

$$a_n = \sum_{i=n}^{\infty} \frac{(i+5)^2}{(i+1)(i+2)(i+3)(i+4)},$$

$$d_n = \frac{n+5}{(n+1)(n+2)(n+4)} = \frac{n+3}{n+5}(a_n - a_{n+1}), \qquad [2.6]$$

$$\hat{T} = \sum_{n=0}^{\infty} \frac{(n+5)^2}{(n+1)(n+2)(n+3)(n+4)},$$

together with

$$V(t) = \begin{cases} -\dfrac{1}{(n+4)!} & \text{if } t \in [a_{n+1}, a_{n+1} + d_n[, \\[2em] -\dfrac{1}{(n+4)!} + \dfrac{F_n}{2} \displaystyle\int_{a_{n+1}+d_n}^{t} \left(\dfrac{s - a_{n+1} - d_n}{a_n - a_{n+1} - d_n} \right) ds, \\[1em] & \text{if } t \in [a_{n+1} + d_n, a_n[. \end{cases} \qquad [2.7]$$

These choices allow the explicit calculation of the following integral

$$U(t) = \int_0^t V(s)ds,$$

the result of which is shown in Figure 2.5.

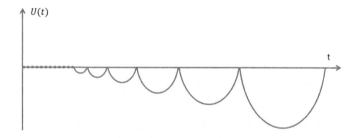

Figure 2.5. *The nontrivial solution for the displacement*

It is then easily checked that function $U(t)$ satisfies:

$$- \left\{ t \in [0, \hat{T}[; \ U(t) = 0 \right\} = \{0\} \cup \{a_n, \ n \in \mathbb{N}^*\}$$

$$- \forall n \in \mathbb{N}, \ \dot{U}^-(a_n) = \frac{1}{(n+3)!}, \ \dot{U}^+(a_n) = -\frac{1}{(n+3)!}, \qquad [2.8]$$

so that $\ddot{U} - F = -2 \displaystyle\sum_{n=0}^{\infty} \frac{\delta_{a_n}}{(n+3)!}.$

Since δ_{a_n} is the Dirac measure, $\ddot{U} - F$ is a negative measure and the function $U(t)$ gives a non-trivial solution of the bouncing problem [2.2].

Two other examples will now be presented in the following sections in order to prove that the ill-posedness of the impact problem is a very general situation. In the first example, the mechanical model is the same as the one studied above except that the perfectly elastic impact is changed into a dissipative impact. In the second example, an elastic stiffness is added to be closer to most of the problems dealt with in the present book.

2.1.2.2. The case with completely inelastic bounces

At first sight, it might seem that the existence of two solutions for the Cauchy problem with smooth data results from the choice $e = 1$ in the impact law since any strictly positive velocity before impact always gives a strictly negative velocity after impact, as seen in equation [2.8]. In order to be sure that the ill-posedness does not result from this constitutive relation, the case $e = 0$, that is the so-called completely inelastic case or in other words the case where the particle always remains at rest in some time interval to the right of an impact time is considered. Once again a counterexample to uniqueness is built, but this time through more intricate calculations than in the case of the elastic impact. This new counterexample, extremely useful for the understanding of the dynamics with impacts, was proposed by Ballard [BAL 00].

To begin with, it is clear that the loading can no longer have a constant sign. In fact, it is easily seen that if the force was always positive and the initial data at rest against the wall as previously, then the only solution to the impact problem would be $U \equiv 0$. Indeed, let F be a positive force and U be a solution to the impact problem with zero initial data, which is not identically equal to zero, therefore passes through strictly negative values at some time. Let t' be such a time where $U(t') < 0$. As $U(t)$ is a continuous function, a time t'' can be defined by $t'' = inf\{t \mid t < t', U(t) < 0\}$ so $U(t'') = 0$. But due to the perfectly inelastic restitution coefficient, we would also have $\dot{U}^+(t'') = 0$, so

$$U(t') = \int_{t''}^{t'} \dot{U}(s)\,ds = \int_{t''}^{t'} \left(\int_{t''}^{s} F(t)\,dt\right) ds \geq 0$$

which contradicts the fact that $U(t') < 0$.

In order to build a $\mathbb{C}^\infty$ loading, the Massin function already used in the case of the elastic impact given by equations [2.3] and [2.4] and matched with zero on a part of each interval is also used for the perfectly inelastic case, but its sign is changed alternatively . This leads to the following function

$$F(0) = 0,$$

$$F(t) = \begin{cases} -f_{1,n}\,\rho\left(\dfrac{t - \dfrac{1}{n+1}}{d_{1,n}}\right), & for\ t \in [\dfrac{1}{n+1}, \dfrac{1}{n+1} + d_{1,n}[, \\[4mm] 0, & for\ t \in [\dfrac{1}{n+1} + d_{1,n}, \dfrac{1}{n} - d_{2,n}[, \\[4mm] f_{2,n}\rho\left(\dfrac{t - \dfrac{1}{n} + d_{2,n}}{d_{2,n}}\right), & for\ t \in [\dfrac{1}{n} - d_{2,n}, \dfrac{1}{n}[, \end{cases} \qquad [2.9]$$

where

$$n \in \mathbb{N}^* \text{ and } (f_{1,n})_{n \in \mathbb{N}^*},\ (f_{2,n})_{n \in \mathbb{N}^*},\ (d_{1,n})_{n \in \mathbb{N}^*},\ (d_{2,n})_{n \in \mathbb{N}^*}$$

are positive sequences, which will be explicitly given later; the function $\rho(.)$ is defined by equation [2.3] given in the previous counterexample, and the partition of each interval satisfies

$$d_{1,n} < \frac{1}{2}\left(\frac{1}{n} - \frac{1}{n+1}\right),\quad d_{2,n} < \frac{1}{2}\left(\frac{1}{n} - \frac{1}{n+1}\right). \qquad [2.10]$$

So finally the loading has the following shape.

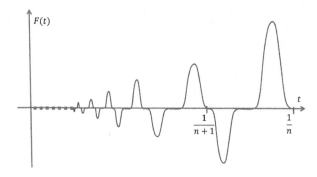

Figure 2.6. *Qualitative shape of the loading $F(t)$*

The sub-sequences $(f_{1,n})_{n\in\mathbb{N}^*}$, $(f_{2,n})_{n\in\mathbb{N}^*}$, $(d_{1,n})_{n\in\mathbb{N}^*}$, $(d_{2,n})_{n\in\mathbb{N}^*}$ are chosen in such a way that the impact problem has two solutions v^I and v^{II}, and that the corresponding functions u^I and u^{II} satisfy:

$$\text{If } n \text{ is even, then} \begin{cases} u^I\left(\frac{1}{n}\right) = 0, \\ v^I\left(\frac{1}{n}\right) = 0, \end{cases} \quad \begin{cases} u^{II}\left(\frac{1}{n}\right) = -u_n, \\ v^{II}\left(\frac{1}{n}\right) = v_n. \end{cases}$$

$$\text{If } n \text{ is odd, then} \begin{cases} u^I\left(\frac{1}{n}\right) = -u_n, \\ v^I\left(\frac{1}{n}\right) = v_n, \end{cases} \quad \begin{cases} u^{II}\left(\frac{1}{n}\right) = 0, \\ v^{II}\left(\frac{1}{n}\right) = 0, \end{cases}$$

[2.11]

where the two sequences $\{u_n\}$ and $\{v_n\}$ are introduced with $f_{i,n}$, $d_{i,n}$, $i = 1,2$ defined as:

$$\begin{cases} \forall n \in \mathbb{N}^*, \quad u_n = \dfrac{1}{n^4 2^n}, \quad v_n = \dfrac{1}{2^n}, \\[2mm] f_{1,n} = \dfrac{n^3}{2^n}, \quad f_{2,n} = f_{1,n}\dfrac{d_{1,n}}{d_{2,n}} + \dfrac{v_n}{d_{2,n}}, \\[2mm] d_{1,n} = \dfrac{1}{2n^3}\left(1 - \sqrt{1 - \dfrac{4n^3}{(n+1)^4}}\right), \quad d_{2,n} = \dfrac{\frac{2n^2}{n+1}d_{1,n} - n^3 d_{1,n}^2 - \frac{2}{n^4}}{1 + n^3 d_{1,n}}. \end{cases}$$

[2.12]

This time $\hat{T}$ is taken equal to $\dfrac{1}{n_0}$ where n_0 is any non-zero integer, the solutions are then given by:

for $n \geq n_0$, $w^I(0) = 0$, $w^{II}(0) = 0$,

$$w^I(t) = \begin{cases} v_{n+1} - f_{1,n}\displaystyle\int_{\frac{1}{n+1}}^{t} \rho\left(\dfrac{s - \frac{1}{n+1}}{d_{1,n}}\right) ds, \quad t \in [\frac{1}{n+1}, \frac{1}{n+1} + d_{1,n}[, \\[3mm] 0, \qquad\qquad\qquad\qquad\qquad\qquad t \in [\frac{1}{n+1} + d_{1,n}, \frac{1}{n}[, \end{cases}$$

$$w^{II}(t) = \begin{cases} -f_{1,n}\displaystyle\int_{\frac{1}{n+1}}^{t} \rho\left(\dfrac{s - \frac{1}{n+1}}{d_{1,n}}\right) ds, \quad t \in [\frac{1}{n+1}, \frac{1}{n+1} + d_{1,n}[, \\[3mm] -f_{1,n}d_{1,n}, \qquad\qquad\qquad\qquad t \in [\frac{1}{n+1} + d_{1,n}, \frac{1}{n} - d_{2,n}[, \\[3mm] -f_{1,n}d_{1,n} + f_{2,n}\displaystyle\int_{\frac{1}{n}-d_{2,n}}^{t} \rho\left(\dfrac{s - \frac{1}{n} + d_{2,n}}{d_{2,n}}\right) ds, \quad t \in [\frac{1}{n} - d_{2,n}, \frac{1}{n}[. \end{cases}$$

[2.13]

Functions $v^I(t)$ and $v^{II}(t)$ should fit equations [2.11], so:

$$t = 0 \begin{cases} v^I(0) = 0, \\ v^{II}(0) = 0, \end{cases}$$

$$\text{if } t \in [\frac{1}{2p+1}, \frac{1}{2p}[, \ (2p \geq n_0), \begin{cases} v^I(t) = w^I(t), \\ v^{II}(t) = w^{II}(t), \end{cases} \qquad [2.14]$$

$$\text{if } t \in [\frac{1}{2p}, \frac{1}{2p-1}[, \ (2p-1 \geq n_0), \begin{cases} v^I(t) = w^{II}(t), \\ v^{II}(t) = w^I(t). \end{cases}$$

Functions $U_1(t)$ and $U_2(t)$ are then built by a simple integration

$$U_1(t) = \int_0^t v^I(s)ds, \quad U_2(t) = \int_0^t v^{II}(s)ds, \qquad [2.15]$$

It can actually be checked that these two functions satisfy all the conditions to be solutions to the one-dimensional impact problem with completely inelastic impact. Their local shapes are shown in Figure 2.7.

2.1.2.3. The case of a mass–spring system

The last example presented here shows that even when the system involves an internal elastic stiffness, uniqueness still does not hold in general. An elementary one-dimensional mass–spring system as the one shown in Figure 2.8 is considered and submitted to an external force f such that $f \in \mathcal{C}^\infty([0, \hat{T}[; \mathbb{R})$. The mechanical problem considered here reads

$$\begin{cases} \text{Find } U \in \text{MMA}([0,T]; \mathbb{R}) \text{ and } R \in \mathcal{M}([0,T]; \mathbb{R}), \text{ such that} \\[2mm] \text{i) } m\ddot{U} + U = F + R, \quad t > 0, \\[2mm] \text{ii) } U(0) = 0, \ \dot{U}^+(0) = 0, \\[2mm] \text{iii) } U \leq 0, \ R \leq 0, \ UR = 0, \\[2mm] \text{iv) } U(t) = 0 \Longrightarrow \dot{U}^+(t) = 0. \end{cases} \qquad [2.16]$$

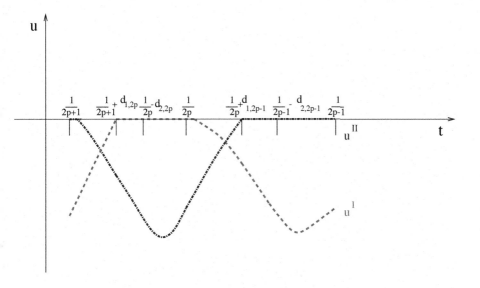

Figure 2.7. *Local shapes of the solutions U_I and U_{II} on interval $[\frac{1}{2p+1}, \frac{1}{2p-1}[$*

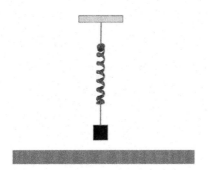

Figure 2.8. *Mass supported by an elastic spring above an obstacle*

As can be seen in system [2.16], the impacts are still assumed to be fully inelastic. The external loading is the same as in the previous example given by [2.9], which is intuitively required by the inelastic impacts.

After calculations that, although relatively tedious, are in fact similar to those of the case of elastic impacts and of the perfectly inelastic case without spring, two solutions can be built to problem [2.16] using a $\mathbb{C}^\infty$ loading. They read as follows where, as previously, $\hat{T} = \dfrac{1}{n_0}$ with $n_0 \in \mathbb{N}^*$.

Then

$$\text{for } n \geq n_0,$$

$$w^I(0) = 0,$$

$$w^{II}(0) = 0,$$

$$w^I(t) = \begin{cases} \sin(t - \dfrac{1}{n+1})u_{n+1} + \cos(t - \dfrac{1}{n+1})v_{n+1} & t \in [\dfrac{1}{n+1}, \dfrac{1}{n+1} + d_{1,n}[, \\ -f_{1,n}\displaystyle\int_{\frac{1}{n+1}}^{t} \cos(t-s)\rho\left(\dfrac{s - \frac{1}{n+1}}{d_{1,n}}\right) ds, & \\ 0, & t \in [\dfrac{1}{n+1} + d_{1,n}, \dfrac{1}{n}[\end{cases} \tag{2.17}$$

$$w^{II}(t) = \begin{cases} -f_{1,n}\displaystyle\int_{\frac{1}{n+1}}^{t} \cos(t-s)\rho\left(\dfrac{s - \frac{1}{n+1}}{d_{1,n}}\right) ds, & t \in [\dfrac{1}{n+1}, \dfrac{1}{n+1} + d_{1,n}[, \\ -f_{1,n}\displaystyle\int_{\frac{1}{n+1}}^{\frac{1}{n+1}+d_{1,n}} \cos(t-s)\rho\left(\dfrac{s - \frac{1}{n+1}}{d_{1,n}}\right) ds, & t \in [\dfrac{1}{n+1} + d_{1,n}, \dfrac{1}{n} - d_{2,n}[, \\ -f_{1,n}\displaystyle\int_{\frac{1}{n+1}}^{\frac{1}{n+1}+d_{1,n}} \cos(t-s)\rho\left(\dfrac{s - \frac{1}{n+1}}{d_{1,n}}\right) ds, & t \in [\dfrac{1}{n} - d_{2,n}, \dfrac{1}{n}[, \\ +f_{2,n}\displaystyle\int_{\frac{1}{n}-d_{2,n}}^{t} \cos(t-s)\rho\left(\dfrac{s - \frac{1}{n} + d_{2,n}}{d_{2,n}}\right) ds. & \end{cases} \tag{2.18}$$

Conditions [2.11] are introduced in the same way as previously to insure that the two solutions are actually different, so:

$$v^I(0) = 0,$$

$$v^{II}(0) = 0,$$

$$\text{if } t \in [\dfrac{1}{2p+1}, \dfrac{1}{2p}[, \ (2p \geq n_0), \qquad \begin{cases} v^I(t) = w^I(t), \\ v^{II}(t) = w^{II}(t), \end{cases} \tag{2.19}$$

$$\text{if } t \in [\dfrac{1}{2p}, \dfrac{1}{2p-1}[, \ (2p-1 \geq n_0), \qquad \begin{cases} v^I(t) = w^{II}(t), \\ v^{II}(t) = w^I(t). \end{cases}$$

The conclusion then follows from

$$U_1(t) = \int_0^t v^I(s)ds, \quad U_2(t) = \int_0^t v^{II}(s)ds. \tag{2.20}$$

It can actually be checked that these two functions satisfy all the conditions to be solutions to the one-dimensional impact problem with completely inelastic impact and elastic restoring force.

2.1.3. *The functional framework for well-posedness*

The existence results briefly recalled in section 2.1.1, in particular the first one given in [SCH 78], hold as soon as the external force is time integrable. But the counterexamples presented in section 2.1.2 show that even if the force is infinitely differentiable, there is no uniqueness, which first suggested that the Cauchy problem is always ill-posed. The idea of increasing the smoothness of the external force up to analytical was due to Danilo Percivale. All the counterexamples to uniqueness are such that $t = 0$ is an accumulation point of zeros of the loading function, this could not be the case for analytical functions who possess a finite number of zero in any bounded interval. Indeed, Percivale proved in [PER 85] that the impact problem studied by Schatzman becomes well-posed if the external force is analytical. The result of uniqueness under an analytical external force is in fact obtained for any finite collection of rigid bodies.

Ballard gave another proof of uniqueness, using classical tools. After obtaining the existence of an analytic solution in an interval to the right of the origin and then assuming that another solution in MMA exists in the same interval, classical estimates show that this solution in MMA is in fact the same as the analytic one (see [BAL 05]).

The statement of problem [2.1] has been extended for n particles in $\mathbb{R}^d$ and by setting $\mathcal{G}(t, U) \equiv \mathcal{F}(U) + F(t)$ the following theorem proved in [CHA 14] gives the optimal well-posedness result.

THEOREM 2.1.– Let us assume that function $\mathcal{G} : [0, T] \times \mathbb{R}^{nd} \longrightarrow \mathbb{R}^{nd}$ satisfies the following hypothesis:

i) the function $U \mapsto \mathcal{G}(t, U)$ satisfies a Lipschitz condition of modulus $\kappa > 0$ for $t \in [0, T]$, i.e.:

$$\forall t \in [0, T], \ \forall \ (U_1, U_2) \in \mathbb{R}^{nd}, \|\mathcal{G}(t, U_1) - \mathcal{G}(t, U_2)\| \leq \kappa \|(U_1 - U_2)\|,$$

where $\|.\|$ stands for the Euclidean norm in $\mathbb{R}^{nd}$.

ii) the function $(t, U) \mapsto \mathcal{G}(t, U)$ is analytic.

Then, problem [2.1] has a unique solution.

The function $\mathcal{G}$ could probably depend also on $\dot{U}$ but as yet the result has not been obtained. The precise functional framework that ensures well-posedness is to be found between infinitely differentiable and analytical. Comments about this statement will be presented along these lines in Chapter 8.

Having found sufficient conditions for the Cauchy problem to be well posed, an approximation of the solution can now be obtained by the use of computational methods.

2.2. Computational methods

This section contains different computational methods that can be used to determine the trajectory of a solution taking into account possible impact times. Let us consider a solution in a given time interval $[0, T]$ containing an impact time τ shown in Figure 2.9. During the phase before the impact time, the trajectory is solution to an ordinary differential equation as is the case for the phase after the impact time. But how can the solution be computed on the whole time interval? Two different approaches can be taken. The first consists of detecting the impact time and giving a condition on the velocity $\dot{u}^+(\tau)$ to compute the phase after the impact. This is called an event-driven method. The second method consists of having a global approach by operating a discretization of the time interval $[0, T]$ and building an algorithm that copes with possible impacts.

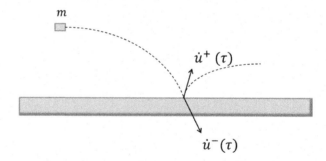

Figure 2.9. *A trajectory may involve time intervals where the velocity varies smoothly and impacts where the velocity jumps*

2.2.1. *The event-driven methods*

2.2.1.1. *The algorithms*

An event-driven method consists of coupling a time integration method of the differential equation with a method enabling to pass from one no contact phase to another. Any time integration method can be used and in some cases explicit analytical solutions can be computed. The crucial point here lies in how the velocity after the impact is computed. The normal component of the velocity after the impact $\dot{u}_n^+(\tau)$ is given by the impact law but no impact law exists for the tangential component. Most event-driven algorithms that can be found in the literature postulate another impact law for the tangential component of the velocity. The determination of the tangential component of the velocity can however be justified. Indeed, an impact time τ is defined by the fact that $u_n(\tau) = 0$ and that $\forall t \in]\tau - \eta, \tau[$ $u_n(t) < 0$ for an $\eta > 0$. When the solution in a no contact phase is known analytically, the impact time τ is obtained by solving the equation $u_n(t) = 0$. When the solution is calculated through a time discretization, a dichotomy method is used to determine τ with a given accuracy. Once this impact time has been determined, the initial conditions needed to solve the differential equation corresponding to the next no contact phase must be computed. The displacement is continuous so the values of the displacement at time τ are $(u_n(\tau) = 0, u_t(\tau))$, but the velocity is of bounded variation and its discontinuities are found at impact times. How is the right velocity $(\dot{u}_n^+(\tau), \dot{u}_t^+(\tau))$ calculated when the left velocity $(\dot{u}_n^-(\tau), \dot{u}_t^-(\tau))$ is known? Reverting to the equations of motion written in the sense of distributions written as:

$$
\left\{
\begin{array}{l}
m\ddot{u}_t = \mathcal{F}_t(u_t, u_n) + F_t + R_t, \\
\\
m\ddot{u}_n = \mathcal{F}_n(u_t, u_n) + F_n + R_n.
\end{array}
\right. \qquad t > 0, \qquad [2.21]
$$

these equations are then integrated over τ and as $\mathcal{F}_t(u_t, u_n)$, $\mathcal{F}_n(u_t, u_n)$, F_t and F_n are continuous functions their integrals over the singleton τ is zero so:

$$
\left\{
\begin{array}{l}
\text{i)} \quad m(\dot{u}_t^+(\tau) - \dot{u}_t^-(\tau)) = \int_\tau R_t \equiv R_t^\tau, \\
\\
\text{ii)} \quad m(\dot{u}_n^+(\tau) - \dot{u}_n^-(\tau)) = \int_\tau R_n \equiv R_n^\tau.
\end{array}
\right. \qquad [2.22]
$$

These two equations are insufficient to determine the four unknowns $\dot{u}_n^+(\tau)$, $\dot{u}_t^+(\tau)$, R_n^τ and R_t^τ. The impact law $\dot{u}_n^+(\tau) = -e\dot{u}_n^-(\tau)$ gives explicitly $\dot{u}_n^+(\tau)$.

Finally, assuming as a constitutive assumption, that the Coulomb friction law still holds between the tangential velocity and the integral of the reaction over the impact time given by equation [2.22], then the pair $(\dot{u}_t^+(\tau), R_t^\tau)$ must belong to the intersection of the graph of equation [2.22(i)], which is a straight line in the $(\dot{u}_t^+(\tau), R_t^\tau)$-plane and the graph [1.1(b)] of the friction law. The following three situations may consequently occur referring to Figure 2.10 for the notations:

– either the straight line D_1 (graph of equation [2.22(i)]) intersects the graph of the Coulomb's law on its vertical part at $(0, R_t^1)$;

– or the straight line D_2 (graph of equation [2.22(i)]) intersects the graph of the Coulomb's law on its left hand side at $(\dot{u}_t^2, -\mu R_n)$;

– or the straight line D_3 intersects the graph of the Coulomb's law on its right hand side at $(\dot{u}_t^3, -\mu R_n)$.

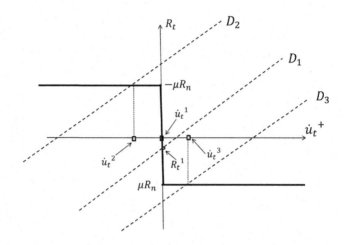

Figure 2.10. *Determining* $(\dot{u}_t^+(\tau), R_t^\tau)$ *at impact time*

These event-driven methods can be very efficient as shall be seen further on but they must be used with the utmost care. In the following section, a very simple example is given where such a method would be unable to determine a solution.

2.2.1.2. *An elementary example where event driven fails*

Let us consider the one-dimensional problem shown in Figure 2.11, which modelizes a ball bouncing on a horizontal obstacle.

The equation of motion is given by:

$$\begin{cases} m\ddot{u} = g \\ u(0) = -5, \quad \dot{u}(0) = 0. \end{cases} \qquad [2.23]$$

The parameters are chosen as follows: $m = 1$, $g = 10$ and the restitution coefficient $e = 0.5$.

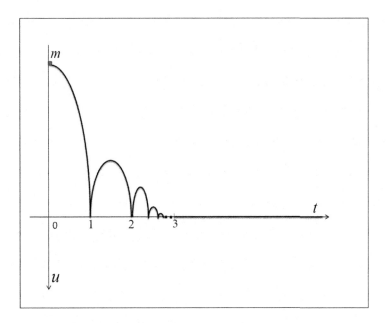

Figure 2.11. *Trajectory of the bouncing ball*

It is elementary to calculate the trajectory of the ball that is given by:

$$\begin{cases} u(t) &=& 5t^2 - 5 & t \in [0,1], \\ u(t) &=& 5t^2 - 15t + 10 & t \in [1,2], \\ u(t) &=& 5t^2 - \dfrac{45}{2}t + 25 & t \in [2,\dfrac{5}{2}], \\ u(t) &=& 5t^2 - 5(6 - \dfrac{3}{2^n})t & t \in [3 - \dfrac{1}{2^{n-1}}, 3 - \dfrac{1}{2^n}], \qquad [2.24] \\ & & +5(3 - \dfrac{1}{2^{n-1}})(3 - \dfrac{1}{2^n}) & \\ u(t) &=& 0 & t \in [3, +\infty[. \end{cases}$$

An event-driven algorithm would not be able to cope with such a trajectory. Indeed, infinitely many impact times would have to be detected before attaining time $t = 3$.

2.2.1.3. *When to use event-driven algorithms*

The simple example given by problem [2.23] shows that any trajectory containing infinitely many impact times could not be computed with an event-driven method. Indeed, such a method would involve infinitely many impact detections, therefore would never solve the problem in finite time. This means that before applying an event-driven method, it is necessary to be sure that the impact times are isolated and that in the time period considered there are a finite number of impact times. In fact, on many occasions in this book, the impact times are found to be isolated.

2.2.2. The time stepping algorithms

2.2.2.1. The algorithms

The time stepping algorithms consist, as their name suggests, of discretizing the time interval $[0, T]$ over which the differential system is to be integrated. A time step is thus defined by $h = T/n$, $t_0 = 0$ and $t_{i+1} = t_i + h$. Bearing in mind that $\dot{u}$ is a function of bounded variation, integrating system [2.21] on the time interval $]t_i, t_{i+1}]$ yields:

$$\begin{cases} m(\dot{u}_t^+(t_{i+1}) - \dot{u}_t^+(t_i)) = \int_{t_i}^{t_{i+1}} (\mathcal{F}_t(u_t, u_n) + F_t)ds + \int_{]t_i, t_{i+1}]} R_t, \\ \\ m(\dot{u}_n^+(t_{i+1}) - \dot{u}_n^+(t_i)) = \int_{t_i}^{t_{i+1}} (\mathcal{F}_n(u_t, u_n) + F_n)ds + \int_{]t_i, t_{i+1}]} R_n. \end{cases} \qquad [2.25]$$

New unknowns are introduced at this point that are coherent with the fact that the reaction is a measure. The vector $\tilde{R}^{i+1} = (\tilde{R}_t^{i+1}, \tilde{R}_n^{i+1})$ is defined by:

$$\tilde{R}_t^{i+1} = \frac{1}{h} \int_{]t_i, t_{i+1}]} R_t \quad \text{and} \quad \tilde{R}_n^{i+1} = \frac{1}{h} \int_{]t_i, t_{i+1}]} R_n.$$

A numerical scheme must then be chosen to approximate $\int_{t_i}^{t_{i+1}} (\mathcal{F}_t(u_t, u_n) + F_t)ds$ and $\int_{t_i}^{t_{i+1}} (\mathcal{F}_n(u_t, u_n) + F_n)ds$. Many choices are available; however, our intention here is not to explore these different choices but to give an overall vision of these schemes. In fact, whatever the numerical integration scheme adopted, system [2.25] can be written as:

$$\dot{U}^{i+1} = V^i + h\mathrm{H}\tilde{R}^{i+1}, \qquad [2.26]$$

where $\dot{U}^{i+1} = (\dot{u}_t^{i+1}, \dot{u}_n^{i+1})$ is an approximation of $(\dot{u}_t^+(t_{i+1}), \dot{u}_n^+(t_{i+1}))$ and the two-dimensional vector $V^i = (V_t^i, V_n^i)$ represents the value of the vector $\dot{U}^{i+1}$ calculated in the case of no contact (i.e. with zero reaction), which is entirely determined by the values of the approximation at t_i and the choice of the integration scheme. The two-by-two real matrix H depends on the choice of the integration scheme. The displacement

$(u_t(t^{i+1}), u_n(t^{i+1}))$ is approximated by (u_t^{i+1}, u_n^{i+1}) given by an integration scheme such as the implicit Euler method, for example:

$$\left\{ \begin{array}{l} u_t^{i+1} = u_t^i + h\dot{u}_t^{i+1}, \\[2mm] u_n^{i+1} = u_n^i + h\dot{u}_n^{i+1}. \end{array} \right. \qquad [2.27]$$

In order to persuade the reader that any numerical discretization can be written as [2.26], a particular numerical integration scheme is chosen, for example the theta method, combined to the implicit Euler method [2.27] and then applied in the linear case where $\mathcal{F}(U) = -KU$, as defined in Chapter 2. Thus, the discretization of [2.25] yields:

$$(m\mathrm{I} + h^2\theta K)\dot{U}^{i+1} = m\dot{U}^i - hKU^i + h((1-\theta)F^i + \theta F^{i+1}) + h\tilde{R}^{i+1}. \qquad [2.28]$$

The two-by-two matrix $m\mathrm{I} + h^2\theta K$ is not singular because its eigenvalues are strictly positive, the matrix K being positive definite. So by setting:

$$\mathrm{H} = (m\mathrm{I} + h^2\theta K)^{-1},$$

and,

$$V^i = (m\mathrm{I} + h^2\theta K)^{-1}(m\dot{U}^i - hKU^i + h((1-\theta)F^i + \theta F^{i+1})),$$

expression [2.28] can be written as [2.26].

The scheme chosen here is an implicit scheme if θ is different from zero. In general, implicit schemes are preferred for nonsmooth dynamical problems. Indeed, such schemes are shown to be able to cope with strong nonlinearities without having to reduce drastically the time step as would be the case for an explicit scheme. This implicit scheme is applied here to the linear case. In the nonlinear case, if an implicit scheme is used, a fixed point iteration must be performed at each time step. It is interesting to stress the fact that as the velocity is known to be of bounded variation, it is useless to use a higher order scheme than the one chosen here. More details concerning the choice of the numerical scheme and its implementation in the nonlinear case are found in [JEA 99].

2.2.2.2. *The implementation*

The full description of the numerical time stepping method can now be given:

– if no contact occurs, the vector $\tilde{R}^{i+1}$ is identically equal to zero so [2.26] provides an approximation of the velocity at t_{i+1};

– but if there is contact then the system [2.26] consists of two equations with four unknowns. The contact conditions and the friction conditions have to be added to system [2.26] in order to determine the four unknowns $\dot{u}_t^{i+1}, \dot{u}_n^{i+1}, \tilde{R}_t^{i+1}, \tilde{R}_n^{i+1}$. These conditions can be written in the following way:

$$\begin{cases} \dot{u}_n^{i+1} = -eV_n^i, \\ \\ (\dot{u}_t^{i+1}, \tilde{R}_t^{i+1}) \text{ belongs to the graph of the Coulomb's law.} \end{cases} \quad [2.29]$$

These contact and friction conditions added to system [2.26] determine explicitly the unknowns $\dot{u}_t^{i+1}, \dot{u}_n^{i+1}, \tilde{R}_t^{i+1}, \tilde{R}_n^{i+1}$ (as in section 3.2.1.1, see Figure 2.10).

The great advantage of this time stepping method is that any number of impacts, finite or infinite, can occur during a time interval. The method can therefore be used without having made sure beforehand that the impact times are isolated, which is a necessary preliminary for event-driven methods.

REMARK 2.1.– The time stepping methods modelized by [2.26] coupled to [2.29] have been shown to converge toward the unique solution of the continuous problem by Monteiro Marquès [MON 93]. In fact, as mentioned in section 3.1.1, that is how Monteiro Marquès proved the existence result.

2.2.3. *When the trajectories remain in contact*

2.2.3.1. *The sliding problem*

There are a certain number of situations where the trajectories are known to remain in contact. This knowledge of the behavior of the trajectories comes from a thorough analysis of the problem that must be performed before any computation takes place. In these situations, the number of variables is reduced and the choice of an event-driven algorithm is natural. When the trajectories remain strictly in contact, u_n is identically equal to zero; there is no contact law and no possibility of impact, so system [2.21] can be rewritten as:

$$\begin{cases} m\ddot{u}_t = \mathcal{F}_t(u_t, 0) + F_t + R_t, \\ 0 = \mathcal{F}_n(u_t, 0) + F_n + R_n, \quad t > 0 \\ \mu R_n \le R_t \le -\mu R_n, \quad \begin{cases} |R_t| < -\mu R_n \implies \dot{u}_t = 0, \\ |R_t| = -\mu R_n \implies \dot{u}_t = -\lambda R_t. \end{cases} \end{cases} \quad [2.30]$$

This system can then be written as two different second-order smooth differential equations depending on whether the motion is sliding to the right or to the left:

– a positive sliding motion implies that the reaction is on the left-hand side of the Coulomb cone so $R_t = \mu R_n$. Inserting this relation into system [2.30], the reaction can be eliminated and the motion is governed by the following equation:

$$m\ddot{u}_t = \mathcal{F}_t(u_t, 0) - \mu \mathcal{F}_n(u_t, 0) + F_t - \mu F_n; \qquad [2.31]$$

– a negative sliding motion implies that the reaction is on the right-hand side of the Coulomb cone so $R_t = -\mu R_n$. Inserting this relation into system [2.30], the reaction can also be eliminated and the motion is governed by the following equation:

$$m\ddot{u}_t = \mathcal{F}_t(u_t, 0) + \mu \mathcal{F}_n(u_t, 0) + F_t + \mu F_n. \qquad [2.32]$$

Two differential equations govern the motion but which one is to be used? Is the mass sliding to the right, to the left or is it motionless? In this section, the external loading (F_t, F_n) shall be supposed constant and the algorithm shall be given in this particular case. Further on the algorithm shall be adapted to the case where the loading is oscillating.

The mathematical results established at the beginning of this chapter ensure the uniqueness of the solution. This implies that if a trajectory satisfying all the equations of the problem is built, then it shall be the unique solution. A technical result that will be essential for the computation of a trajectory that remains in contact is detailed here.

LEMMA 2.1.– When the tangential velocity under constant loading becomes equal to zero after a sliding phase, either the velocity stays equal to zero for all time or the velocity changes sign (i.e. the sliding direction changes).

PROOF.– Assume the tangential velocity becomes zero at some time $t = t^*$, then the normal component of the reaction $R_n(t^*) = -F_n - \mathcal{F}_n(u_t(t^*), 0)$ is either the same as the one of an equilibrium solution or not:

– the existence and the structure of the set of equilibrium states of problem [2.1] will be investigated in the following chapters. Nevertheless, it can be formally stated that if the normal component of the reaction at time t^*, say $R_n(t^*)$ is the same as the one of an equilibrium state, then there is a jump of the tangential component of the reaction and the unique solution of problem [2.30] is given by the trajectory obtained up to the time t^* when the tangential velocity reaches zero followed by the constant function $(u_t(t), u_n(t)) = (u_t(t^*), 0)$ for $t > t^*$;

– if the normal component of the reaction at time t^*, $R_n(t^*)$ is not the same as the one of an equilibrium state, then $(u_t(t^*), 0)$ is not an equilibrium solution. It is known, due to [CHA 14], that there can be no jump in the tangential velocity for the sliding problem, so this tangential velocity is continuous at time t^*. It cannot be equal to zero because $(u_t(t^*), 0)$ is not an equilibrium solution, therefore the velocity for $t > t^*$ is either strictly positive or strictly negative (indeed an accumulation of zeros to the right of t^* is excluded by the analyticity result). Let us suppose that for an interval of time on the left of t^*, the velocity was positive (we could of course assume that the velocity was negative and the argument would easily be transposed). If the velocity stays positive for an interval of time to the right of t^*, then the sliding continues on the same side of the cone, therefore u_t is the solution of a classical ordinary differential equation, equation [2.31] in this case, and there can be no discontinuity of the second derivative. There is consequently a minimum of the function $\dot{u}_t$ at $t = t^*$ and the second derivative of u_t is therefore equal to zero at $t = t^*$. But that case would give:

$$m\ddot{u}_t(t^*) = 0 = \mathcal{F}_t(u_t(t^*), 0) + F_t + R_t(t^*), \atop 0 = \mathcal{F}_n(u_t(t^*), 0) + F_n + R_n(t^*), \qquad t \geq t^*,$$

which would imply that $(u_t(t^*), 0)$ is an equilibrium solution, which contradicts the fact that we are in the case where $R_n(t^*)$ does not belong to the set of R_n, which correspond to equilibrium solutions.

The velocity must therefore be strictly negative for an interval of time to the right of t^*, so the velocity changes sign and the sliding continues on the other side of the cone. This implies that for $t > t^*$, equation [2.31] no longer governs the movement but that equation [2.32] does. $\qquad\square$

Of course to compute the trajectory of a mass remaining in contact, initial conditions must be added to the equations [2.30]. These initial conditions, in the case of trajectories remaining in contact, must inevitably correspond to conditions in contact. The initial conditions are written as:

$$u_n(0) = 0, \quad \dot{u}_n(0) = 0, \quad u_t(0) = u_0, \quad \dot{u}_t(0) = v_0.$$

If v_0 is strictly positive, the trajectory is obtained by solving the second-order differential equation [2.31] with the two initial conditions (u_0, v_0). If v_0 is strictly negative, the trajectory is obtained by solving the second-order differential equation [2.32] with the two initial conditions (u_0, v_0). As long as the velocity does not change sign, these trajectories are computed either analytically or by a numerical scheme, and when the velocity becomes equal to zero lemma 2.1 determines the future trajectory. However, if v_0 is equal to zero, either $(u_0, 0)$ is an equilibrium solution and the mass stays motionless, or the initial data $(u_0, 0)$ is not an equilibrium solution and a complementary information is needed to establish whether the mass shall start sliding to the right or to the left. Further on in the book, this question shall be studied in detail. Indeed, the topology of the sets of equilibrium solutions plays an important part in the answer to this question.

2.2.3.2. *Description of the algorithm*

The computation of a trajectory remaining in contact consists of alternatively integrating two second-order differential equations. In fact, the only difficulty lies in determining which differential equation must be integrated and the time at which the change of differential equation occurs.

Let us suppose that there has been a sliding phase and that τ is such that $\dot{u}_t(\tau) = 0$. Then, the trajectory for $t > \tau$ shall be determined by:

1) Compute the normal component of the reaction at time τ:

$$R_n(\tau) = -\mathcal{F}_n(u_t(\tau), 0) - F_n;$$

2) If the normal component of the reaction $R_n(\tau)$ belongs to the set of normal reactions corresponding to equilibrium solutions, then the mass stays motionless from then on eventually after a jump in the tangential component of the reaction.

3) If this normal component of the reaction does not belong to the set of normal reactions corresponding to equilibrium solutions, then:

i) if $R_t^-(\tau) = -\mu R_n(\tau)$, there shall be a jump in the tangential component of the reaction so $R_t^+(\tau) = \mu R_n(\tau)$. A positive sliding is initiated and the trajectory shall be solution to [2.31] until the mass stops sliding at time $\bar{\tau}$ where $\dot{u}_t(\bar{\tau}) = 0$. Return to (1) setting $\tau = \bar{\tau}$;

ii) if $R_t^-(\tau) = \mu R_n(\tau)$ there shall be a jump in the tangential component of the reaction so $R_t^+(\tau) = -\mu R_n(\tau)$. A negative sliding is initiated and the trajectory shall be solution to [2.32] until the mass stops sliding at time $\bar{\tau}$ where $\dot{u}_t(\bar{\tau}) = 0$. Return to (1) setting $\tau = \bar{\tau}$.

This algorithm shall be widely used to compute the solutions that are given in this book. The integration of the differential equations shall be explicit in the linear case, whereas a fourth-order Runge–Kutta scheme shall be used when $\mathcal{F}$ is not linear.

The Equilibrium States

In this chapter, the equilibrium states are determined and then classified. Whether in the case where the restoring force is linear or in the general large strains case, the methods applied to find the equilibrium states are identical. The method consists of solving the equilibrium equations and then determining in the first place under which conditions the normal components of the solutions of the equilibrium equations are strictly negative, i.e. are not in contact, and then, in the second place, under which conditions the solutions of the equilibrium equations in contact, that is such that the normal component of the displacement is equal to zero, satisfy the Coulomb friction conditions.

3.1. In the linearized case

The method presented here can be applied to any system with a linear restoring force. It is discussed in the following in the case of the simple mass–spring system introduced in Chapter 2 and recalled in Figure 3.1.

3.1.1. The equilibrium states

The equilibrium states must satisfy the following equilibrium equations:

$$\begin{cases} K_t u_t + W u_n = F_t + R_t \\ \\ W u_t + K_n u_n = F_n + R_n, \end{cases} \qquad [3.1]$$

where (F_t, F_n) denotes the external force and $K = \begin{pmatrix} K_t & W \\ W & K_n \end{pmatrix}$ the stiffness matrix of the system of springs. But they must also satisfy the non-regularized unilateral contact and Coulomb friction laws, which reduce in statics to

$$\begin{cases} u_n \leq 0, \; R_n \leq 0, \; u_n.R_n = 0, \\[2mm] |R_t| \leq \mu|R_n|. \end{cases} \qquad [3.2]$$

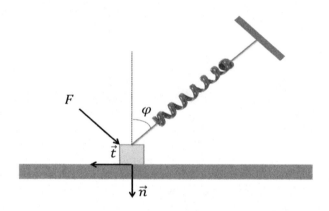

Figure 3.1. *The mass–spring model*

The equilibrium states are now determined by separating the two cases, equilibrium solutions without contact or equilibrium solutions in contact:

1) A solution without contact is such that the reaction is identically equal to zero so that by [3.1] a no contact equilibrium solution satisfies:

$$\begin{cases} u_n = \dfrac{A}{detK}, \quad u_t = \dfrac{K_n F_t - W F_n}{detK} \\[3mm] R_n = R_t = 0, \end{cases} \qquad [3.3]$$

where the quantity A is defined as $A = K_t F_n - W F_t$. Conditions [3.2] then imply that [3.3] will be an equilibrium solution only if $A \leq 0$.

2) For equilibrium solutions in contact with the obstacle, the normal component of the displacement is equal to zero. So that by introducing $u_n = 0$ into systems [3.1] and [3.2], equilibrium solutions must verify

$$\begin{cases} K_t u_t = F_t + R_t \\ W u_t = F_n + R_n \\ |R_t| \leq \mu|R_n|, \end{cases} \qquad [3.4]$$

which yields the following determination of equilibrium solutions in contact

$$\begin{cases} R_t = \dfrac{K_t}{W} R_n + \dfrac{A}{W} \\[2mm] |R_t| \leq \mu |R_n|. \end{cases} \qquad\qquad [3.5]$$

3.1.2. *Classification of the equilibria*

The set of equilibrium solutions depends strongly on the parameter A, which is qualitatively interesting since A depends only on the data and A is, up to the strictly positive constant $det K$, the solution u_n of the associated unconstrained problem 3.1. The equilibrium solutions can be deduced from Figures 3.2–3.4. The thin continuous lines are the boundaries of Coulomb's cone; eventual dashed lines delimit the range of admissible values for R_n; on each graph, the thick continuous line is the straight line of equation $R_t = \frac{K_t}{W} R_n + \frac{A}{W}$. According to the previous formula, the set of equilibrium solutions is then given by the intersection of this thick straight line with Coulomb's cone in addition to possible equilibrium solutions without contact. This implies that in some cases eventual bounds for R_n are determined. Such a graphic discussion was used in [BAS 03] and [MAR 94] for the quasi-static evolution.

In the first case, as shown in Figure 3.2, the loads and the stiffness parameters are such that $A \equiv K_t F_n - W F_t$ is strictly negative. In this case, the thick line intersects the R_n axis for some positive value of R_n that is out of the cone. Then, either mass m does not touch the obstacle, that means the unique solution is without contact ($\mu \leq K_t/W$), or there is one equilibrium solution without contact and infinitely many others in contact, one of which is in impending slip ($\mu > K_t/W$).

In the next case (Figure 3.3, $A \equiv K_t F_n - W F_t = 0$), the thick line crosses the boundary of Coulomb's cone exactly at the vertex. Then, either the vertex, that is impending sliding with no reaction, is the unique equilibrium solution ($\mu < K_t/W$), or all the points of the left boundary of the cone are equilibrium solutions, that is there are infinitely many solutions in impending sliding, one of which being with no reaction ($\mu = K_t/W$), or there are infinitely many equilibrium solutions in strictly stuck contact (the thick open half line) in addition to a solution in contact with no reaction (at the vertex of the cone), which is the unique solution in impending sliding ($\mu > K_t/W$).

Figure 3.4 shows the third case ($A \equiv K_t F_n - W F_t > 0$), which involves bounds on the admissible normal reactions.

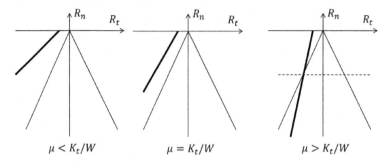

Figure 3.2. *The set of equilibrium solutions in the* $\{R_t, R_n\}$ *plane when* $A < 0$

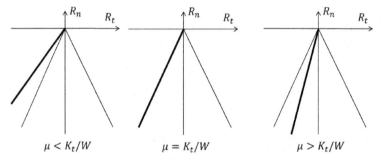

Figure 3.3. *The set of equilibrium solutions in the* $\{R_t, R_n\}$ *plane when* $A = 0$

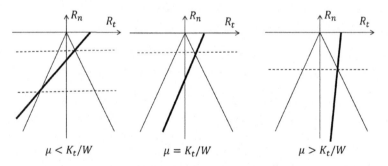

Figure 3.4. *The set of equilibrium solutions in the* $\{R_t, R_n\}$ *plane when* $A > 0$

The dependence of the set of equilibria on the stiffness parameters, friction coefficient and external loads is summarized in Table 4.1.

$\mu < \dfrac{K_T}{W}$	$\mu = \dfrac{K_T}{W}$	$\mu > \dfrac{K_T}{W}$
$A < 0$ One solution without contact	solution without contac	One solution without contact + One solution in impending positive slip + infinitely many solutions in strict stuck contact
$A = 0$ One solution in grazing contact	One solution in grazing contact + infinitely many solutions in impending positive slip	One solution in grazing contact + infinitely many solutions in strict stuck contact
$A > 0$ Two solutions in impending positive and negative slip + infinitely many solutions in strict stuck contact	One solution in impending negative slip + infinitely many solutions in strict stuck contact	One solution in impending negative slip + infinitely many solutions in strict stuck contact

Table 3.1. *The equilibrium states with respect to parameters A and μ*

The classification given in Table 3.1 brings to light the strong coupling between normal and tangential components. Indeed, referring to the first column of the table, if the quantity $A \equiv K_t F_n - W F_t$ is strictly positive for a given value of the external loading, infinitely many equilibrium solutions in contact exist. However, if the tangential component of the external force is increased sufficiently for A to become negative, only a single equilibrium out of contact remains.

3.2. Equilibria at large strains

In this section, the mass–spring system shown in Figure 3.5 is the same as in the previous section but the restoring forces now allow large displacements:

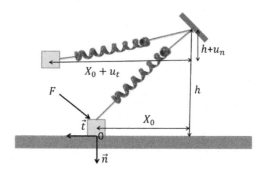

Figure 3.5. *Mass–spring system with nonlinear strain*

The equilibrium states must satisfy the following system:

$$\begin{cases} \mathcal{N}_t(u_t, u_n) + F_t + R_t = 0, \\ \mathcal{N}_n(u_t, u_n) + F_n + R_n = 0, \\ u_n < 0, \quad R_n < 0 \ \text{ and } \ u_n R_n = 0, \\ |R_t| \leq -\mu R_n. \end{cases} \quad [3.6]$$

Determining and classifying the equilibrium solutions becomes considerably harder than in the linear case. The method, however, is the same as in the linear case. Conditions ensuring the existence of equilibrium solutions out of contact are sought for first and then the existence of equilibrium solutions in contact is tackled.

3.2.1. *Equilibrium solutions out of contact*

If equilibrium solutions out of contact exist for some external data k, h, X_0, F_t, F_n, the reaction (R_t, R_n) must be equal to zero so that these solutions must satisfy the following system:

$$\begin{cases} \mathcal{N}_t(u_t, u_n) + F_t = 0, \\ \mathcal{N}_n(u_t, u_n) + F_n = 0, \\ u_n < 0. \end{cases} \quad [3.7]$$

REMARK 3.1.– In fact, the unilateral contact condition amounts to checking that eventual solutions of the two first algebraic equations of system [3.7] are compatible with the single inequality $u_n < 0$, in which case the Coulomb friction law would be automatically satisfied by $\{R_t, R_n\} = \{0, 0\}$, as in the case of the linear restoring force.

Using the expression of the nonlinear strains given in Chapter 2, system [3.7] can be rewritten into the explicit form:

$$
\begin{cases}
(X_0 + u_t) \left[1 - \dfrac{\sqrt{X_0^2 + h^2}}{\sqrt{(X_0 + u_t)^2 + (h + u_n)^2}} \right] = \dfrac{F_t}{k}, \\[3mm]
(h + u_n) \left[1 - \dfrac{\sqrt{X_0^2 + h^2}}{\sqrt{(X_0 + u_t)^2 + (h + u_n)^2}} \right] = \dfrac{F_n}{k}, \\[3mm]
u_n < 0.
\end{cases}
\qquad [3.8]
$$

As mentioned in the previous section, solutions to the two equations of system [3.8] are computed, then the conditions under which these solutions are not in contact are exhibited. However, this time the algebraic system with unknowns u_t and u_n is relatively intricate so that the case $F_n \neq 0$ shall be looked into first, and the particular case $F_n = 0$ shall be studied later.

3.2.1.1. *The normal component F_n is not equal to zero*

An elementary manipulation gives in this case

$$
u_t + X_0 = \frac{F_t}{F_n}(u_n + h).
$$

Introducing this into system [3.8]:

$$
u_n + h - \frac{F_n}{k} = \sqrt{\frac{X_0^2 + h^2}{(h + u_n)^2 \left(1 + \left(\dfrac{F_t}{F_n} \right)^2 \right)}}(h + u_n),
$$

so that

$$
\left(u_n + h - \frac{F_n}{k} \right)^2 = h^2 \alpha(F_t, F_n)^2.
\qquad [3.9]
$$

In expression [3.9], the real parameter $\alpha(F_t, F_n)$ depends on the geometry and is given by

$$\alpha(F_t, F_n) = \sqrt{\left(1 + \left(\frac{X_0}{h}\right)^2\right) \Big/ \left(1 + \left(\frac{F_t}{F_n}\right)^2\right)}.$$

Equation [3.9] has two solutions:

$$u_n = \frac{F_n}{k} - h + h\alpha(F_t, F_n) \quad \text{and} \quad u_n = \frac{F_n}{k} - h - h\alpha(F_t, F_n).$$

These solutions shall be solutions to system [3.8] if and only if they are strictly negative.

The quantities h and $\alpha(F_t, F_n)$ being strictly positive, the second solution is strictly smaller than the first one. So that if the first solution is negative, then the second one shall also be negative.

The conditions implying the existence of equilibrium solutions out of contact are given by:

$$\begin{cases} \dfrac{F_n}{kh} - 1 + \alpha(F_t, F_n) < 0 \implies & \text{two solutions out of contact,} \\[4mm] \begin{cases} \dfrac{F_n}{kh} - 1 + \alpha(F_t, F_n) > 0 \\ \quad \text{and} \\ \dfrac{F_n}{kh} - 1 - \alpha(F_t, F_n) < 0 \end{cases} \implies & \text{one solution out of contact,} \quad [3.10] \\[4mm] \dfrac{F_n}{kh} - 1 - \alpha(F_t, F_n) > 0 \implies & \text{no solution out of contact.} \end{cases}$$

These conditions define three zones in the $\left\{\frac{F_t}{kh}, \frac{F_n}{kh}\right\}$ plane as shown in Figure 3.6: zone 1 (no shading) where no solutions out of contact exist, zone 2 (light gray) where a single solution out of contact exists and zone 3 (dark gray) where two solutions out of contact coexist. The origin has a special status: infinitely many equilibrium solutions exist.

REMARK 3.2.– Two solutions can be interpreted very simply since the spring is linearly elastic and the kinematics allows large displacements: one of the solutions corresponds to an extension of the spring and the other to a traction, both under the same force in the direction of the spring.

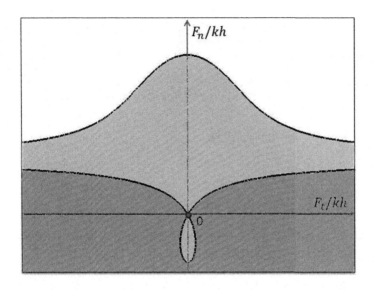

Figure 3.6. *Values of the force corresponding to stationary solutions out of contact*

REMARK 3.3.– The direction of the force is given by F_t/F_n and the quantity X_0/h gives the direction of the reference configuration of the spring so that $F_t/F_n = X_0/h$ implies $\alpha(F_t, F_n) = 1$ (in other words, if the direction of the force is the direction of the reference configuration of the spring, then $\alpha(F_t, F_n) = 1$ and the spring extends in the direction of its reference configuration).

3.2.1.2. *The normal component F_n is equal to zero*

When the normal component of the force is equal to zero, the previous computations cannot be carried out and the system reduces to:

– if $F_t \neq 0$, system [3.8] immediately gives the trivial solutions:

$$\begin{cases} u_n = -h \\ u_t = \dfrac{F_t}{k} - X_0 + h\beta, \end{cases} \qquad [3.11]$$

and

$$\begin{cases} u_n = -h \\ u_t = \dfrac{F_t}{k} - X_0 - h\beta. \end{cases} \qquad [3.12]$$

The geometrical dimensionless parameter $\beta = \sqrt{1 + \left(\dfrac{X_0}{h}\right)^2}$ has been introduced and shall often appear in the following.

It is interesting to observe that as $F_n = 0$, this solution corresponds to a horizontal position of the spring at the level $-h$, which is quite trivial since only a horizontal force is applied:

– if $F_t = 0$, then system [3.8] gives

$$1 - \frac{\sqrt{X_0^2 + h^2}}{\sqrt{(X_0 + u_t)^2 + (h + u_n)^2}} = 0,$$

which is the equation of a circle of center $(-X_0, -h)$ and radius $\sqrt{X_0^2 + h^2}$ (the natural length of the spring), so that the trivial result is that in the absence of external forces all the points belonging both to the circle of center $(-X_0, -h)$ and radius $\sqrt{X_0^2 + h^2}$ and to the open negative half-plane are equilibrium solutions. This obviously results from the fact that our simple model has only one spring, so that the result would be slightly modified in the case of several springs (as in Figure 1.4(a), for example).

3.2.2. Stationary solutions in contact

The contact and friction conditions imply that an additional parameter appears, which is the friction coefficient μ. In order to find solutions in contact, $u_n = 0$ must be inserted into system [3.6] so that the remaining unknowns are u_t, R_t, and R_n. The equilibrium problem is now given by:

$$\begin{cases} (X_0 + u_t)\left[1 - \dfrac{h\beta}{\sqrt{(X_0 + u_t)^2 + h^2}}\right] = \dfrac{F_t + R_t}{k}, \\[4mm] \left[1 - \dfrac{h\beta}{\sqrt{(X_0 + u_t)^2 + h^2}}\right] = \dfrac{F_n + R_n}{kh}, \\[4mm] R_n \leq 0, \ |R_t| \leq -\mu R_n, \end{cases}$$

[3.13]

which can be rewritten as:

$$\begin{cases} \text{i)} \ (X_0 + u_t)(F_n + R_n) = h(F_t + R_t), \\[4mm] \text{ii)} \ 1 - \dfrac{h\beta}{\sqrt{(X_0 + u_t)^2 + h^2}} = \dfrac{F_n + R_n}{kh}, \\[4mm] \text{iii)} \ R_n \leq 0, \ |R_t| \leq -\mu R_n. \end{cases}$$

[3.14]

Equations [3.14] imply that $0 < kh - (F_n + R_n) < kh\beta$. So that the normal component of the reaction is bounded and satisfies:

$$kh(1 - \beta) - F_n < R_n < kh - F_n.$$

Equation [3.14(ii)] can be solved analytically, giving u_t as a function of R_n:

$$u_t(R_n) = -X_0 \pm h\sqrt{\frac{k^2 h^2 \beta^2}{(F_n + R_n - kh)^2} - 1}. \tag{3.15}$$

Inserting the expression of u_t obtained in [3.15] into equation [3.14(i)], the equilibrium equation is given by the following expression that defines a curve in the $\{R_t, R_n\}$ plane:

$$(R_t + F_t)^2 = (F_n + R_n)^2 \left(\frac{k^2 h^2 \beta^2}{(F_n + R_n - kh)^2} - 1 \right). \tag{3.16}$$

Finding equilibrium solutions in contact then amounts to finding the intersection of this curve [3.16] with the Coulomb cone given by [3.14(iii)].

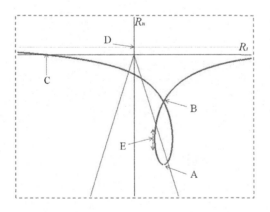

Figure 3.7. *An example of curve [3.16]*

3.2.2.1. *The external force is not equal to zero*

The case where the external force is not equal to zero can be taken as the generic case. The equilibrium solutions shall first be classified in this case. An example of curve [3.16] is shown in Figure 3.7 in the $\{R_t, R_n\}$ plane, where five particular points, denoted by A, B, C, D and E, have been added. The coordinates of these points are:

$$
\left\{
\begin{array}{l}
A \left|
\begin{array}{l}
-F_t \\
kh(1 - \beta - \dfrac{F_n}{kh}),
\end{array}
\right.
\quad
B \left|
\begin{array}{l}
-F_t \\
-F_n,
\end{array}
\right. \\[2em]
C \left|
\begin{array}{l}
-F_t - F_n \sqrt{\dfrac{\beta^2}{(\dfrac{F_n}{kh} - 1)^2} - 1} \\[1.5em]
0,
\end{array}
\right. \\[3em]
D \left|
\begin{array}{l}
0 \\
kh(1 - \dfrac{F_n}{kh}),
\end{array}
\right.
\quad
E \left|
\begin{array}{l}
-kh(\dfrac{F_t}{kh} + (\beta^{2/3} - 1)^{3/2}) \\[1em]
-kh(\dfrac{F_n}{kh} - 1 + \beta^{2/3}).
\end{array}
\right.
\end{array}
\right.
\qquad [3.17]
$$

REMARK 3.4.– In the case of linearized strains studied in the previous section, the curve representing the equilibrium equation given by formula [3.16] was simply a straight line given by [3.5].

Modifying the values of F_t and F_n has the effect of translating the curve in the $\{R_t, R_n\}$ plane, so that F_t and F_n will consequently be the parameters of the following discussion for given values of X_0, h and μ. Moreover, modifying the values of X_0, h and μ does not lead to qualitative changes of the curve shown in Figure 3.7. When the intersection of the curve with the Coulomb cone is empty, there are no equilibrium solutions and conversely a non-empty intersection corresponds to the existence of equilibrium solutions. So that the values of the parameters corresponding to the existence of equilibrium solutions can be determined. This is shown in Figures 3.8–3.12:

$$\boxed{- \frac{F_n}{kh} \leq 1 - \beta}$$: no solutions in contact

Point A is in the positive half-plane so that there are no solutions in contact, whatever the value of μ or F_t when $\dfrac{F_n}{kh} < 1 - \beta$. This situation is shown in Figure 3.8. When $\dfrac{F_n}{kh} = 1 - \beta$, there are no solutions in contact except for $F_t = 0$ when there is a single equilibrium in grazing contact because the point A is at the vertex of the cone.

$$\boxed{- \frac{F_n}{kh} \geq 1}$$: infinitely many equilibrium solutions in contact

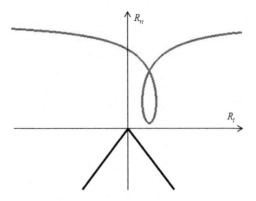

Figure 3.8. *Non-existence of equilibrium solutions in contact*

Point D (which corresponds to the intersection of the horizontal asymptote with the R_n axis) is in the negative half-plane so that the entire curve is in the negative half-plane. In this case, there always exists a part of the curve of non-zero length that is strictly inside the cone, so that there always exist infinitely many equilibrium solutions in contact. This situation is shown in Figure 3.9.

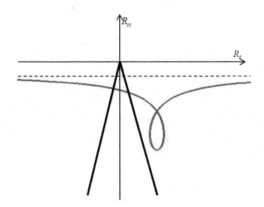

Figure 3.9. *Existence of equilibrium solutions in contact for* $\dfrac{F_n}{kh} \geq 1$

$$-\left|1 - \beta < \frac{F_n}{kh} < 1\right|:$$

In this range, the curve intersects the R_t axis at two distinct points (at only one point when $F_n = 0$). If these two points are either side of the origin, then the curve intersects the Coulomb cone whatever the value of the friction coefficient μ (see

Figure 3.10). The values of R_t corresponding to the intersection of the curve with the R_t axis satisfy:

$$(R_t + F_t)^2 = F_n^2 \left(\frac{k^2 h^2 \beta^2}{(F_n - kh)^2} - 1 \right).$$

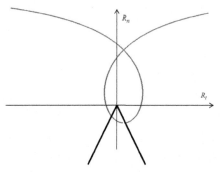
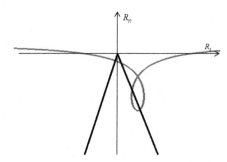

a) Intersection either side of the origin when $F_n < 0$

b) Intersection either side of the origin when $F_n > 0$

Figure 3.10. *Unconditional existence of equilibrium states in the case*
$$\frac{F_n}{kh} - 1 + \alpha(F_t, F_n) > 0$$

So that the curve intersects the R_t axis either side of the origin if and only if:

$$F_t^2 - F_n^2 \left(\frac{k^2 h^2 \beta^2}{(F_n - kh)^2} - 1 \right) < 0.$$

Using the parameter $\alpha(F_t, F_n)$ introduced earlier, this condition can be written as:

$$-\alpha(F_t, F_n) < \frac{F_n}{kh} - 1 < \alpha(F_t, F_n).$$

As $\dfrac{F_n}{kh}$ is strictly smaller than 1, the second inequality is automatically satisfied, so that when $\dfrac{F_n}{kh} - 1 + \alpha(F_t, F_n) > 0$, unconditional existence of equilibrium states is obtained.

When the two intersections of the curve with the R_t axis are on the same side of the origin, the existence of equilibrium solutions is conditioned in general by the value of the friction coefficient as can be seen in Figure 3.11. There are however some values of the parameters where the two intersections of the curve with the R_t axis are on the same side of the origin but where the curve interects the Coulomb cone for any value of the friction coefficient μ. This particular case is shown in Figure 3.12. It corresponds to the case where the two intersections of the curve with the R_t axis are on the same side of the origin but where the curve also intersects the R_n axis. This condition can be given by determining the values of the parameters for which the point E is on the opposite side of the R_n axis to the side where the curve intersects the R_t axis. So that using the coordinates of point E given in [3.17], the following condition is obtained:

$$\frac{F_n}{kh} > 1 - \beta^{2/3} \quad \text{and} \quad \frac{|F_t|}{kh} < (\beta^{2/3} - 1)^{3/2}.$$

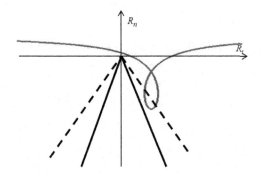

Figure 3.11. *Conditional existence of equilibrium solutions in contact for $\dfrac{F_n}{kh} < 1$: thick continuous border of the cone: $\mu < \mu_c$, dashed border of the cone $\mu > \mu_c$*

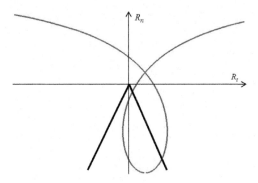

Figure 3.12. *Unconditional existence of equilibrium solutions in contact when the curve intersects the R_n axis*

When equilibrium solutions in contact exist, the generic situation is the existence of infinitely many solutions in contact. However, there may exist a unique solution in contact, for example as stated above, when $\dfrac{F_n}{kh} = 1 - \beta$, $F_t = 0$ (when the minimum of the equilibrium curve is at the vertex of the Coulomb cone) or whenever the intersection of the curve and the cone is reduced to the single point where the curve is tangent to a border of the cone.

To summarize the above discussion, three different zones of the $\left\{ \dfrac{F_t}{kh}, \dfrac{F_n}{kh} \right\}$ plane are shown in Figure 3.13: Zone 1 (light gray) where no equilibrium solution in contact exist, zone 2 (gray) where equilibrium solutions in contact exist when the coefficient of friction μ is sufficiently large and zone 3 (dark gray) where equilibrium solutions in contact exist whatever the value of μ.

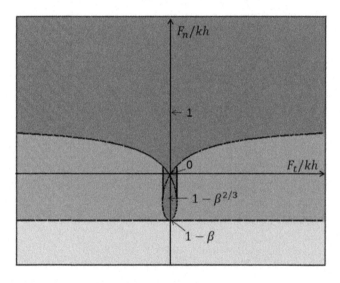

Figure 3.13. *Values of the force corresponding to equilibrium solutions in contact*

3.2.2.2. No external force

The particular case where no external force is applied to the system is now explored. As stated at the end of section 4.2.1, the components of the equilibrium solutions out of contact when there is no external force are found on the part of the circle of center $(-X_0, -h)$ and radius $\sqrt{X_0^2 + h^2}$, which is in the negative half-plane. The two extremities of this arc $((0, 0)$ and $(0, -2X_0))$ are therefore in grazing contact. When there is no external force, point B is at the origin so that there exist infinitely many equilibrium states, as shown in Figure 3.14, in the $\{R_t, R_n\}$ plane. The part of the curve in the $\{R_t, R_n\}$ plane, which is in the Coulomb cone with a non-zero reaction, corresponds to tangential positions, which fill a closed interval included in the segment $[0, -2X_0]$ whose length depends on the value of μ. The set of equilibria in contact (therefore with $u_n = 0$) without any external force is given as follows:

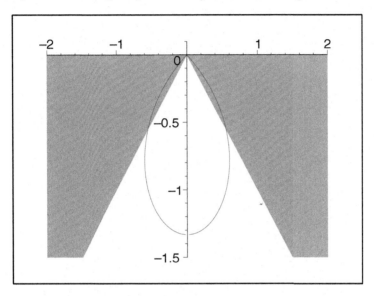

Figure 3.14. *The equilibria strictly in contact in the $\{R_t, R_n\}$ plane without external forces*

– for $\mu < X_0/h$, the tangential components of the equilibria consist of all the points in the closed interval $[-X_0 + h\mu, -X_0 - h\mu]$ strictly included in the interval $(0, -2X_0)$, and of the points $u_t = 0$ and $u_t = -2X_0$;

– as μ increases, the length of the interval of equilibria $[-X_0 + h\mu, -X_0 - h\mu]$ increases until $\mu = X_0/h$ where the set of equilibria consists of the whole closed interval $[0, -2X_0]$. It is immediate that this interval cannot be extended by continuing to increase μ since any point outside this interval would correspond to an extended length of the spring, which is not possible without external force;

– observe that in the limit case $\mu = 0$, the whole curve shown in Figure 3.14 shrinks to the single point $\{R_t = 0, R_n = kh(1 - \beta)\}$, so that in this limit case there are only three equilibrium positions: $(u_t = 0, u_n = 0), (u_t = -X_0, u_n = 0)$ and $(u_t = -2X_0, u_n = 0)$. The spring is compressed at the point $(u_n = 0, u_t = -X_0)$, which obviously means instability in the case without friction, while the spring is at rest in the two other positions. The set of equilibria of the SD oscillator, which does not involve friction conditions, is thus recovered [CAO 06].

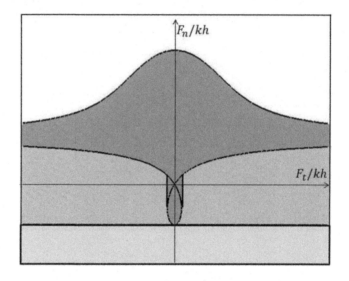

Figure 3.15. *The set of equilibrium solutions*

3.2.2.3. *Equilibrium states with respect to the loading*

Whenever there are no equilibrium solutions in contact, there exist equilibrium solutions out of contact, and conversely whenever there are no equilibrium solutions out of contact, there exist equilibrium solutions in contact. Moreover, equilibrium solutions in contact and out of contact may exist simultaneously, as can be seen by superimposing Figures 3.6 and 3.13. Figure 3.15 summarizes the previous results when the components of the external force are not equal to zero. Four different situations appear in the $\left\{ \dfrac{F_t}{kh}, \dfrac{F_n}{kh} \right\}$ plane: there is a first zone (light gray) where no equilibrium solution in contact exists, a second zone (no shading) where no equilibrium solution out of contact exists, a third zone (dark gray) where there is coexistence of equilibrium solutions out of contact and in contact for any value of the friction coefficient μ and finally a fourth zone (medium gray) where there is coexistence of equilibrium solutions out of contact and in contact as long as μ is sufficiently large.

4

Stability

This chapter concerns the stability of the equilibria in the linear case, which was first explored by direct numerical computations; however, analytical calculations often give more elegant and complete results. These results are to be found in the first section of this chapter. However, the corresponding stability results, although complete, remain in some sense ill suited to systems involving Coulomb friction, so a new notion of stability is introduced, specially adapted to systems with unilateral contact and Coulomb friction. Whereas stability results in mechanics classically concern perturbations of the initial data in the phase space, this new notion of stability concerns the robustness of the dynamics with respect to perturbations of the external forces. In this context, a stability conjecture is stated, and backed up first by analytical computations in the case of simple models and then by numerical computations for more complex or larger size systems. All this analysis is carried out in the case of the linear restoring force, that is when $\mathcal{F}(U) = -KU$.

4.1. Stability of the equilibrium states in the classical sense

The main qualitative results of Chapter 4 were the following:

– The structure of the set of equilibria depends only on the sign of two parameters which are $A := K_t F_n - W F_t$ and $\mu - \frac{K_t}{W}$.

– Equilibrium states out of contact always exist when $A < 0$. The equilibrium state without contact is then unique. Moreover, if $\mu - \frac{K_t}{W} > 0$, then the equilibrium state without contact coexists with infinitely many equilibrium states in contact, which therefore have a strictly negative normal component of the reaction.

– If $A \geq 0$, there are no equilibrium states out of contact. Moreover, if $A = 0$ and $\mu - \frac{K_t}{W} < 0$ the set of equilibria reduces to a single state in grazing contact.

– If $A > 0$, there always exist infinitely many equilibrium states in contact, which all have a strictly negative normal component of the reaction. If in addition $\mu - \frac{K_t}{W} \geq 0$, then this set fills completely a half-line in the $\{R_t, R_n\}$ plane, while it fills only a bounded interval if $\mu - \frac{K_t}{W} < 0$.

When the restoring force is linear, the dependence on the parameter of the set of equilibrium states has been investigated in the case where the external force F is constant. So on the one hand the set of equilibria is completely and explicitly known, and on the other hand the theoretical results ensure that there exists a single trajectory as soon as initial data compatible with the unilateral conditions are given. A classical stability analysis can therefore be performed.

4.1.1. *Some qualitative phases of the dynamics*

The stability analysis is divided into two main parts. This first part contains the analysis of separate phases of abstract trajectories, and shall be used as technical lemmas in the second part. The second part deals with the stability properties of the equilibrium states. In both parts, some partial results are obtained by direct analytical calculations. These calculations are possible because each phase of the motion is smooth and linear due to the assumption of small strains, and because the successive phases are smoothly matched due to the global regularity of the solution in the case of a constant force. In other cases, these analytical tools, although possible, become too intricate and it will be easier to perform estimates on the iterates of the corresponding discrete dynamical system, and to conclude using the convergence results. This stability analysis is performed by perturbing the equilibria obtained with given forces by initial data out of equilibrium and studying the trajectories without changing the forces.

LEMMA 4.1.– Let $(U^{\mathrm{eq}}, R^{\mathrm{eq}})$ be an equilibrium state with $u_n^{\mathrm{eq}} = 0$ and let $v_{0n} < 0$ be a perturbation of this equilibrium induced by a normal velocity at $t = 0$. Then, the trajectory will involve an impact time $t^{\mathrm{imp}} > 0$.

PROOF.– After a perturbation $v_{0n} < 0$, there exists a time interval in which the particle moves without contact, since the solution is analytical. During this phase, the evolution is described by the following smooth system:

$$\begin{cases} m\ddot{u}_t + K_t u_t + W u_n = F_t, \\ m\ddot{u}_n + W u_t + K_n u_n = F_n, \\ u_t(0) = u_t^{eq} = \dfrac{F_n + R_n^{eq}}{W}, \ u_n(0) = 0, \\ \dot{u}_t(0) = 0, \ \dot{u}_n(0) = v_{0n}, \end{cases} \qquad [4.1]$$

and the solution reads as follows:

$$
\left\{
\begin{aligned}
u_t(t) &= (a_1 \cos(\alpha_1 t) + b_1 \sin(\alpha_1 t))\xi_1 + (a_2 \cos(\alpha_2 t) + b_2 \sin(\alpha_2 t))\xi_2 \\
&\quad + \frac{K_n F_t - W F_n}{K_t K_n - W^2}, \\
u_n(t) &= a_1 \cos(\alpha_1 t) + b_1 \sin(\alpha_1 t) + a_2 \cos(\alpha_2 t) + b_2 \sin(\alpha_2 t) \\
&\quad + \frac{A}{K_t K_n - W^2},
\end{aligned}
\right.
\tag{4.2}
$$

where the coefficients α_i and ξ_i (for $i = 1, 2$) are given by

$$
\begin{aligned}
\alpha_1 &= \sqrt{\frac{(K_t + K_n) + \sqrt{(K_n - K_t)^2 + 4W^2}}{2m}}, \\
\alpha_2 &= \sqrt{\frac{(K_t + K_n) - \sqrt{(K_n - K_t)^2 + 4W^2}}{2m}}, \\
\xi_1 &= \frac{(K_n - K_t) + \sqrt{(K_n - K_t)^2 + 4W^2}}{2W}, \\
\xi_2 &= \frac{(K_n - K_t) - \sqrt{(K_n - K_t)^2 + 4W^2}}{2W}.
\end{aligned}
\tag{4.3}
$$

The parameters a_1, a_2, b_1 and b_2 are determined through the initial conditions and are equal to:

$$
a_1 = \frac{1}{\xi_1 - \xi_2} \left(\frac{A}{K_t K_n - W^2}(\xi_2 + \frac{K_n}{W}) + \frac{R_n^{eq}}{W} \right),
$$

$$
a_2 = \frac{1}{\xi_1 - \xi_2} \left(\frac{A}{K_t K_n - W^2}(\xi_1 + \frac{K_n}{W}) + \frac{R_n^{eq}}{W} \right),
$$

$$
b_1 = \frac{-v_{0n}\xi_2}{\alpha_1(\xi_1 - \xi_2)}, \qquad b_2 = \frac{v_{0n}\xi_1}{\alpha_2(\xi_1 - \xi_2)}.
$$

The proof then follows immediately whether the ratio $\dfrac{\alpha_1}{\alpha_2}$ is rational or not. $\qquad\square$

This result also holds for other cases such as an initial data in grazing contact and a perturbation by a strictly positive tangential velocity, which leads to the following corollary:

COROLLARY 4.1.– Assume that there exists a time for which the reaction is at the vertex of the cone at $t = 0$. Then this time is followed by a phase without contact, followed in turn by a time of impact $t^{\text{imp}} > 0$.

Direct analytical computations establish the following result:

LEMMA 4.2.– Assume that $A = 0$ and $\mu > \dfrac{K_t}{W}$.

Let v_{0t} be a perturbation of the equilibrium in grazing contact induced by a positive tangential velocity at $t = 0$ and let $t^{\text{imp}} > 0$ be the following impact time:

i) if $\dot{u}_t^- (t^{\text{imp}}) \leq 0$ there exists $\hat{t} > t^{\text{imp}}$ with $\dot{u}_t(\hat{t}) = 0$ and $R_n(\hat{t}) < 0$;

ii) if $\dot{u}_t^- (t^{\text{imp}}) > 0$ then there exists $\eta > 0$ such that $\forall v_{0t} \in]0, \eta[$ there exists $\hat{t} > t^{\text{imp}}$ with $\dot{u}_t(\hat{t}) = 0$ and $R_n(\hat{t}) < 0$.

Another important intermediate result deals with the amplitude of the sliding motion. It can be stated as:

LEMMA 4.3.– Assume A is strictly positive and $\mu < \frac{K_t}{W}$, and let t_1 be an initial time such that:

$$\begin{cases} u_t(t_1) = u_{0t}, \quad u_n(t_1) = 0, \\ R_t(t_1) = \mu R_n(t_1) \neq 0 \\ \dot{u}_t(t_1) = 0, \quad \dot{u}_n(t_1) = 0. \end{cases}$$

Then, the trajectory issued from this initial data is such that, if t_2 and t_3 are defined by:

$$\begin{cases} \dot{u}_t(t_2) = 0 \text{ where } u_t(t_2) \text{ is not an equilibrium,} \\ \dot{u}_t(t_3) = 0, \end{cases}$$

then

$$u_t(t_3) > u_t(t_1).$$

The qualitative meaning of lemma 4.3 is shown in Figure 4.1.

PROOF.– From now onwards, the following notations will be used for the sake of simplification:

$$\sqrt{\frac{K_t - \mu W}{m}} = \omega_\alpha, \quad \sqrt{\frac{K_t + \mu W}{m}} = \omega_\beta. \qquad [4.4]$$

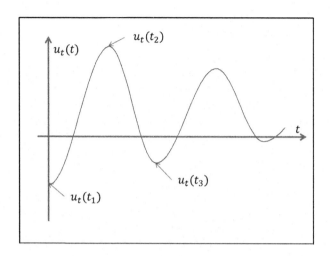

Figure 4.1. *Graphic interpretation of lemma 4.3*

Direct analytical calculations give

$$u_t(t) = \left(u_{0t} - \frac{F_t - \mu F_n}{K_t - \mu W}\right)\cos\omega_\alpha(t - t_1) + \frac{F_t - \mu F_n}{K_t - \mu W} \qquad [4.5]$$

in an interval on the right of t_1.

According to equation [4.5], there exists a time $t_2 > 0$ such that $\dot{u}_t(t_2) = 0$ and $u_t(t_2) = \left(-u_{0t} + 2\frac{F_t - \mu F_n}{K_t - \mu W}\right)$. At this time, the normal reaction is equal to $R_n(t_2) = -Wu_{0t} - \frac{2A}{K_t - \mu W} + F_n$. Assuming that $u_t(t_2)$ is not an equilibrium, by lemma 2.1 the sliding holds now in the negative direction and the evolution is governed by a bilateral process where the initial conditions are

$$u_t(t_2) = -u_{t0} + 2\frac{F_t - \mu F_n}{K_t - \mu W}, \quad \dot{u}_t(t_2) = 0,$$

so

$$u_t(t) = \left(-u_{0t} + 2\frac{F_t - \mu F_n}{K_t - \mu W} - \frac{F_t + \mu F_n}{K_t + \mu W}\right)\cos\omega_\beta(t - t_2) + \frac{F_t + \mu F_n}{K_t + \mu W}.$$

This value of $u_t(t)$ guarantees the existence of a time $t_3 > t_2$ such that $\dot{u}_t(t_3) = 0$. At this time, the tangential displacement is

$$u_t(t_3) = u_{0t} - 2\frac{F_t - \mu F_n}{K_t - \mu W} + 2\frac{F_t + \mu F_n}{K_t + \mu W} = u_t(t_1) + \frac{2\mu A}{K_t^2 - (\mu W)^2}.$$

Since A is strictly positive and $\mu < K_t/W$, it follows that $u_t(t_1) < u_t(t_3)$.

These calculations can straightforwardly be extended to any further times t_i, $i = 3, \ldots n$. In the same way, the initial sliding direction could be taken as negative in which case $u_t(t_3)$ would have been found strictly smaller than $u_t(t_1)$. This lemma shows that the amplitude of the oscillations shall diminish strictly as long as an equilibrium has not been reached. $\square$

4.1.2. The stability results

The stability of all the equilibria described in Chapter 4 can now be obtained. Stability, referred to as "stability in the classical sense," amounts to studying the time evolution of the distance between an equilibrium state and the solution of the Cauchy problem [2.1] where the initial conditions are given in a neighborhood of the equilibrium. If there exists, in such a neighborhood, initial data such that the dynamics diverges from the equilibrium, in a finite time or asymptotically in time, the equilibrium will be unstable. On the other hand, if no perturbation leading to a divergence exists, the equilibrium will be Lyapunov or asymptotically stable.

THEOREM 4.1.– Any equilibrium state without contact is Lyapunov stable.

Without contact, the particle is not subjected to unilateral constraints. This stability result is therefore nothing other than the classical result for the solution of a linear ordinary differential equation.

THEOREM 4.2.– The equilibrium in grazing contact characterized by $A = 0$ and $\mu < K_t/W$ is asymptotically stable.

PROOF.– Although it is relatively long, the proof of this result can be obtained completely analytically, and many other results could be obtained through a similar proof. It consists of four steps.

Step 1: Preliminary

Without restriction, a perturbation induced by an initial velocity is considered. Moreover, based on lemma 4.2, the evolution is identical whether it is induced by a positive tangential velocity or by a negative normal velocity. In this first step, it is shown that the study of the perturbation induced by a negative tangential velocity includes the study of the perturbation by a positive tangential velocity. Indeed, it is proved that after a negative perturbation V_{0t}, there exists a time t_1^{vertex} with $\dot{u}_t(t_1^{\text{vertex}}) > 0$. Thus, after t_1^{vertex}, the procedure shall be the same as that used to determine the evolution after the equilibrium at the vertex of the cone has been perturbed by a positive tangential velocity.

After a perturbation induced by a negative tangential velocity, the evolution is given by a bilateral problem with initial data

$$u_t(0) = \frac{F_n}{W}, \quad \dot{u}_t(0) = V_{0t} < 0.$$

As A is equal to zero, the evolution is given by

$$u_t(t) = V_{0t}\frac{1}{\omega_\beta}\sin\omega_\beta t + \frac{F_t + \mu F_n}{K_t + \mu W}.$$

Let $\hat{t} = \dfrac{\pi}{2\omega_\beta}$ be the time at which $\dot{u}_t(\hat{t}) = 0$. At this time, the particle is not at equilibrium, so the phase of negative slip is followed by a phase of positive slip. During this second phase, the evolution is again described by a bilateral sliding problem but now the initial data are:

$$u_t(\hat{t}) = \frac{V_{0t}}{\omega_\beta} + \frac{F_t + \mu F_n}{K_t + \mu W}, \quad \dot{u}_t(\hat{t}) = 0,$$

so the solution reads

$$u_t(t) = V_{0t}\frac{1}{\omega_\beta}\cos\omega_\alpha(t - \hat{t}) + \frac{F_t - \mu F_n}{K_t - \mu W}.$$

Therefore, there exists a time t_1^{vertex} at which the particle reaches the vertex of the cone, with $u_t(t_1^{\text{vertex}}) = F_n/W$. Since $A = K_t F_n - W F_t = 0$, this time is given by

$$t_1^{\text{vertex}} = \hat{t} + \frac{\pi}{2\omega_\alpha} = \frac{\pi}{2\omega_\beta} + \frac{\pi}{2\omega_\alpha},$$

and then

$$\dot{u}_t(t_1^{\text{vertex}}) = -V_{0t}\frac{\omega_\alpha}{\omega_\beta} > 0,$$

which is the end of the first step.

Step 2: The evolution of the particle includes a series of slip phases and of phases without contact

More specifically, let t_i^{vertex} be a time where the reaction reaches the vertex of the cone with a strictly positive velocity, as calculated in Step 1. Then, the evolution contains a part without contact, a time of impact t_i^{imp}, a new sliding part, and reaches the vertex of the cone once more at t_{i+1}^{vertex}.

After t_i^{vertex}, the trajectory is given by lemma 4.1 until the time of impact t_i^{imp}, so

$$u_t(t_i^{\text{imp}}) = \dot{u}_t(t_i^{\text{vertex}})\frac{\sin\alpha_1(t_i^{\text{imp}} - t_i^{\text{vertex}})}{\alpha_1} + \frac{F_n}{W},$$

$$u_n(t_i^{\text{imp}}) = \frac{\dot{u}_t(t_i^{\text{vertex}})}{\xi_1 - \xi_2}\left(\frac{\sin\alpha_1(t_i^{\text{imp}} - t_i^{\text{vertex}})}{\alpha_1} - \frac{\sin\alpha_2(t_i^{\text{imp}} - t_i^{\text{vertex}})}{\alpha_2}\right) = 0,$$

$$\dot{u}_t^-(t_i^{\text{imp}}) = \frac{\dot{u}_t(t_i^{\text{vertex}})}{\xi_1 - \xi_2}\left(\xi_1\cos\alpha_1(t_i^{\text{imp}} - t_i^{\text{vertex}}) - \xi_2\cos\alpha_2(t_i^{\text{imp}} - t_i^{\text{vertex}})\right),$$

$$\dot{u}_n^-(t_i^{\text{imp}}) = \frac{\dot{u}_t(t_i^{\text{vertex}})}{\xi_1 - \xi_2}\left(\cos\alpha_1(t_i^{\text{imp}} - t_i^{\text{vertex}}) - \cos\alpha_2(t_i^{\text{imp}} - t_i^{\text{vertex}})\right).$$

[4.6]

The velocity after impact is determined by equations [2.22]. Choosing here the restitution coefficient e equal to zero, the normal velocity after impact is equal to zero so that the normal reaction at impact R_n is equal to $-m\dot{u}_n^-(t_i^{\text{imp}})$ and the tangential velocity after the impact is given by

$$\begin{cases} \dot{u}_t^+(t_i^{\text{imp}}) = 0 \ \text{ if } \ |\dot{u}_t^-(t_i^{\text{imp}})| \le \left|\frac{\mu R_n}{m}\right|, \\[2em] \dot{u}_t^+(t_i^{\text{imp}}) = \dot{u}_t^-(t_i^{\text{imp}}) + \frac{\mu R_n}{m} \\[1em] \qquad = \dot{u}_t^-(t_i^{\text{imp}}) - \mu\dot{u}_n^-(t_i^{\text{imp}}) \ \text{ if } \ \dot{u}_t^-(t_i^{\text{imp}}) > \mu\dot{u}_n^-(t_i^{\text{imp}}), \\[2em] \dot{u}_t^+(t_i^{\text{imp}}) = \dot{u}_t^-(t_i^{\text{imp}}) - \frac{\mu R_n}{m} \\[1em] \qquad = \dot{u}_t^-(t_i^{\text{imp}}) + \mu\dot{u}_n^-(t_i^{\text{imp}}) \ \text{ if } \ \dot{u}_t^-(t_i^{\text{imp}}) < -\mu\dot{u}_n^-(t_i^{\text{imp}}). \end{cases}$$

[4.7]

At this point, it is essential to notice that the length of time between t_i^{vertex} and t_i^{imp} is determined by the parameters α_1 and α_2 and does not depend on i. Indeed, the quantity $t_i^{\text{imp}} - t_i^{\text{vertex}}$ is obtained by solving the following equation given in [4.6]:

$$\frac{\sin\alpha_1(t_i^{\text{imp}} - t_i^{\text{vertex}})}{\alpha_1} - \frac{\sin\alpha_2(t_i^{\text{imp}} - t_i^{\text{vertex}})}{\alpha_2} = 0.$$

From now on, the following notation is adopted:

$$t_i^{\text{imp}} - t_i^{\text{vertex}} \equiv \Delta.$$

In the same way, the sign of the tangential velocity before impact, therefore also after impact, is given by the sign of the quantity

$$\frac{(\xi_1 \cos \alpha_1 \Delta - \xi_2 \cos \alpha_2 \Delta)}{\xi_1 - \xi_2},$$

and this quantity depends uniquely on the mechanical parameters of the problem, which define α_1, α_2, ξ_1 and ξ_2 given in [4.3].

In the case where $\dot{u}_t^+(t_i^{\mathrm{imp}}) \geq 0$, the tangential slip will be in the positive direction with initial data equal to $u_t(t_i^{\mathrm{imp}})$ and $\dot{u}_t^+(t_i^{\mathrm{imp}})$. The motion is therefore given by:

$$u_t(t) = \frac{F_t - \mu F_n}{K_t - \mu W} + \dot{u}_t(t_i^{\mathrm{vertex}}) \frac{\sin \alpha_1 \Delta}{\alpha_1} \cos \omega_\alpha (t - t_i^{\mathrm{imp}})$$
$$+ \frac{\dot{u}_t^+(t_i^{\mathrm{imp}})}{\omega_\alpha} \sin \omega_\alpha (t - t_i^{\mathrm{imp}}),$$

and the particle reaches the vertex of the cone at $t_{i+1}^{\mathrm{vertex}}$ when

$$u_t(t_{i+1}^{\mathrm{vertex}}) = \frac{F_n}{W}.$$

Consequently, as $A = 0$ implies that $\dfrac{F_t - \mu F_n}{K_t - \mu W} - \dfrac{F_n}{W} = 0$, the time when the particle reaches the vertex of the cone is given by

$$t_{i+1}^{\mathrm{vertex}} = t_i^{\mathrm{imp}} + \frac{1}{\omega_\alpha} \arctan \left(-\frac{\omega_\alpha \dot{u}_t(t_i^{\mathrm{vertex}}) \sin \alpha_1 \Delta}{\alpha_1 \dot{u}_t^+(t_i^{\mathrm{imp}})} \right).$$

In the case where $\dot{u}_t^+(t_i^{\mathrm{imp}}) < 0$, the tangential slip shall be initiated in the negative direction. The particle shall then stop sliding and, as the only equilibrium solution is the vertex of the cone, the particle shall subsequently start sliding in the positive direction until it reaches the vertex of the cone.

The trajectory consists of a series of phases without contact between time t_i^{vertex} and time t_i^{imp}, followed by slip phases between time t_i^{imp} and $t_{i+1}^{\mathrm{vertex}}$.

Step 3: Let t_i^{vertex} and $t_{i+1}^{\mathrm{vertex}}$ be two consecutive times such that $u_t(t_j^{\mathrm{vertex}}) = F_n/W$, for $j = i, i+1$, then $\dot{u}_t(t_{i+1}^{\mathrm{vertex}}) < \dot{u}_t(t_i^{\mathrm{vertex}})$

In the simple case where $\dot{u}_t^+(t_i^{\mathrm{imp}}) = 0$, the calculations of step 2 yield:

$$\dot{u}_t(t_{i+1}^{\mathrm{vertex}}) = -\frac{\omega_\alpha}{\alpha_1} \dot{u}_t(t_i^{\mathrm{vertex}}) \sin \alpha_1 \Delta,$$

so

$$\frac{\dot{u}_t(t_{i+1}^{\mathrm{vertex}})}{\dot{u}_t(t_i^{\mathrm{vertex}})} \leq \frac{\omega_\alpha}{\alpha_1} < 1.$$

When the sliding is positive after impact, the calculations of step 2 yield:

$$\dot{u}_t(t_{i+1}^{\text{vertex}}) = \omega_\alpha \sqrt{\dot{u}_t^2(t_i^{\text{vertex}}) \left(\frac{\sin \alpha_1 \Delta}{\alpha_1}\right)^2 + \left(\frac{\dot{u}_t^+(t_i^{\text{imp}})}{\omega_\alpha}\right)^2}.$$

And through [4.6] and [4.7], the tangential velocity after impact is written:

$$\dot{u}_t^+(t_i^{\text{imp}}) = \frac{\dot{u}_t(t_i^{\text{vertex}})}{\xi_1 - \xi_2} \left((\xi_1 - \mu)\cos \alpha_1 \Delta - (\xi_2 - \mu)\cos \alpha_2 \Delta\right),$$

finally,

$$\frac{\dot{u}_t(t_{i+1}^{\text{vertex}})}{\dot{u}_t(t_i^{\text{vertex}})} = \lambda,$$

where λ is a constant depending only on the stiffness parameters and on the friction coefficient given by:

$$\lambda = \sqrt{\left(\frac{\omega_\alpha \sin \alpha_1 \Delta}{\alpha_1}\right)^2 + \left(\frac{(\xi_1 - \mu)\cos \alpha_1 \Delta - (\xi_2 - \mu)\cos \alpha_2 \Delta}{\xi_1 - \xi_2}\right)^2}.$$

This constant λ shall be shown to be strictly smaller than one. Indeed, the normal velocity before impact must be positive for the impact to occur so that $\cos \alpha_2 \Delta \leq \cos \alpha_1 \Delta$. The quantity $(\xi_2 - \mu)$ is negative since from equations [4.3] ξ_2 itself is negative, so

$$-(\xi_2 - \mu)\cos \alpha_2 \Delta \leq -(\xi_2 - \mu)\cos \alpha_1 \Delta,$$

and therefore,

$$0 \leq (\xi_1 - \mu)\cos \alpha_1 \Delta - (\xi_2 - \mu)\cos \alpha_2 \Delta \leq (\xi_1 - \xi_2)\cos \alpha_1 \Delta.$$

Finally:

$$\lambda \leq \sqrt{\left(\frac{\omega_\alpha \sin \alpha_1 \Delta}{\alpha_1}\right)^2 + (\cos \alpha_1 \Delta)^2},$$

and,

$$\lambda < \sqrt{(\sin \alpha_1 \Delta)^2 + (\cos \alpha_1 \Delta)^2},$$

so that parameter λ is strictly smaller than 1.

When the sliding is negative after impact, a similar computation establishes that:

$$\frac{\dot{u}_t(t_{i+1}^{\text{vertex}})}{\dot{u}_t(t_i^{\text{vertex}})} = \lambda',$$

where

$$\lambda' = \sqrt{\left(\frac{\omega_\alpha \sin \alpha_1 \Delta}{\alpha_1}\right)^2 + \left(\omega_\alpha \frac{(\xi_1 + \mu) \cos \alpha_1 \Delta - (\xi_2 + \mu) \cos \alpha_2 \Delta}{\omega_\beta (\xi_1 - \xi_2)}\right)^2}.$$

It is then shown that

$$\lambda' < 1.$$

Step 4: Let t_{i+1}^{vertex} be a time such that $u_t(t_{i+1}^{\text{vertex}}) = F_n/W$ where i is the index of a cycle (phase without contact, slip phase). Then,

$$\lim_{i \longrightarrow \infty} \dot{u}_t(t_{i+1}^{\text{vertex}}) = 0.$$

According to Step 3,

$$\dot{u}_t(t_{i+1}^{\text{vertex}}) = \lambda \dot{u}_t(t_i^{\text{vertex}}).$$

So

$$\dot{u}_t(t_{i+1}^{\text{vertex}}) = (\lambda)^i \, \dot{u}_t(t_1^{\text{vertex}}).$$

and as $\lambda < 1$

$$\lim_{i \longrightarrow \infty} \dot{u}_t(t_{i+1}^{\text{vertex}}) = 0. \qquad \qquad \square$$

The stability properties of the three following sets of equilibria are obtained in a similar way. They are stated as follows.

THEOREM 4.3.– The equilibrium in grazing contact ($U^{gr} = (0, F_n/W), R^{gr} = 0$) characterized by $A = 0$ and $\mu = K_t/W$ is Lyapunov stable.

THEOREM 4.4.– All the equilibrium states in impending positive slip characterized by $A = 0$, $\mu = K_t/W$ and a strictly negative reaction are unstable.

THEOREM 4.5.– Let $A \geq 0$ and μ be such that $\mu > K_t/W$ when $A = 0$ or $\mu > 0$ when $A > 0$. Then, all the equilibrium states are Lyapunov stable.

The proof of the stability result stated in theorem 4.6 is slightly different.

THEOREM 4.6.– Let A be strictly negative and μ be such that $\mu > K_t/W$. Then, in addition to the Lyapunov stability of the single equilibrium out of contact, as given by theorem 4.1:

i) the single equilibrium state in imminent sliding is unstable;

ii) all the strictly stuck equilibria are Lyapunov stable.

PROOF.– When A is strictly negative and μ is strictly larger than K_t/W, the equilibrium solution in imminent sliding to the right, given by $u_t^* = \frac{F_t - \mu F_n}{K_t - \mu W}$, has been shown to be unstable. This instability result has been obtained by two different ways in [BAS 06] and [MAR 94]. But there also exist infinitely many equilibrium solutions, which are strictly stuck. The tangential component of the position of any of these ones can be written as $u_t^* - \eta$ where η is strictly positive. Assume such an equilibrium solution is perturbed by a tangential velocity v_0:

– If $v_0 < 0$, then a sliding on the right-hand side of the cone shall occur and the movement shall be solution to the following differential equation:

$$
\begin{cases}
\ddot{u}_t + (K_t + \mu W)u_t = F_t + \mu F_n, \\[2mm]
u_t(0) = u_t^* - \eta, \quad \dot{u}_t(0) = v_0.
\end{cases}
$$

The solution of this equation is

$$
u_t(t) = \frac{F_t + \mu F_n}{K_t + \mu W} - \left(\frac{2\mu A}{K_t^2 - \mu^2 W^2} + \eta\right)\cos\omega_\beta t + \frac{v_0}{\omega_\beta}\sin\omega_\beta t.
$$

When the movement stops at time $\bar{t}$ the tangential component is equal to:

$$
u_t(\bar{t}) = \frac{F_t + \mu F_n}{K_t + \mu W} - \sqrt{\left(\frac{2\mu A}{K_t^2 - \mu^2 W^2} + \eta\right)^2 + (v_0/(\omega_\beta)^2}.
$$

So

$$
u_t(\bar{t}) - (u_t^* - \eta) = \frac{2\mu A}{K_t^2 - \mu^2 W^2} + \eta - \sqrt{\left(\frac{2\mu A}{K_t^2 - \mu^2 W^2} + \eta\right)^2 + (v_0/(\omega_\beta)^2} < 0,
$$

and

$$
|u_t(\bar{t}) - (u_t^* - \eta)| = \left| \sqrt{\left(\frac{2\mu A}{K_t^2 - \mu^2 W^2} + \eta\right)^2 + (v_0/\omega_\beta)^2} \right.
$$

$$
\left. - \sqrt{\left(\frac{2\mu A}{K_t^2 - \mu^2 W^2} + \eta\right)^2} \right|,
$$

$$
|u_t(\bar{t}) - (u_t^* - \eta)| \leq \frac{1}{2\left(\frac{2\mu A}{K_t^2 - \mu^2 W^2} + \eta\right)}|(v_0/\omega_\beta)^2|.
$$

The movement shall stop because $u_t(\bar{t})$ being strictly smaller than $u_t^* - \eta$, which in turn is strictly smaller than u_t^*, is a strictly stuck equilibrium solution and the distance to $u_t^* - \eta$ shall decrease with v_0.

– If $v_0 > 0$, then a sliding on the left-hand side of the cone shall occur and the movement shall be solution to the following differential equation:

$$\begin{cases} \ddot{u}_t + (K_t - \mu W)u_t = F_t - \mu F_n, \\[2mm] u_t(0) = u_t^* - \eta, \quad \dot{u}_t(0) = v_0. \end{cases}$$

The solution of this equation is

$$u_t(t) = u_t^* - \eta \cosh(\sqrt{\mu W - K_t}\,t) + \frac{v_0}{\sqrt{\mu W - K_t}} \sinh(\sqrt{\mu W - K_t}\,t),$$

so that if $0 < v_0 < \eta\sqrt{\mu W - K_t}$ the movement shall stop at time $\bar{t}$ defined by

$$\bar{t} = \operatorname{argtanh}\left(\frac{v_0}{\eta\sqrt{\mu W - K_t}}\right)/\sqrt{\mu W - K_t}.$$

Therefore

$$u_t(\bar{t}) - u_t^* = -\eta\sqrt{1 - \frac{v_0^2}{\eta^2(\mu W - K_t)}} < 0,$$

and finally the following expression is obtained:

$$|u_t(\bar{t}) - (u_t^* - \eta)| \leq \frac{\dfrac{v_0^2}{(\mu W - K_t)}}{2\sqrt{\eta^2 - \dfrac{v_0^2}{(\mu W - K_t)}}}.$$

So that the movement shall stop because $u_t(\bar{t})$ being strictly smaller than u_t^* is a strictly stuck equilibrium solution. Therefore, for sufficiently small values of v_0, $u_t(\bar{t})$ shall be a strictly stuck equilibrium solution whose distance to $u_t^* - \eta$ shall decrease with v_0. $\qquad\square$

The dependence of the set of equilibria on the parameters has been summarized in Table 4.1. Their stability is presented in Table 4.1.

It may be useful in the understanding of the above results to keep in mind the representation of the dynamics in the $\{R_t, R_n\}$ plane, for example the Lyapunov stability of a strictly stuck equilibrium state can be represented by a trajectory jumping from one border of the cone to the other as shown in Figure 4.2. The stability properties shown in this section are illustrated in Figures 4.3–4.5, by a few trajectories obtained numerically by the time stepping method given in Chapter 3.

	$\mu < \dfrac{K_T}{W}$	$\mu = \dfrac{K_T}{W}$	$\mu > \dfrac{K_T}{W}$
$A < 0$	One Lyapunov stable solution	One Lyapunov stable solution	One Lyapunov stable solution + One unstable solution + infinitely many Lyapunov stable solutions
$A = 0$	One asymptotically stable solution	One Lyapunov stable solution + infinitely many unstable solutions	Infinitely many Lyapunov stable solutions
$A > 0$	Infinitely many Lyapunov stable solutions	Infinitely many Lyapunov stable solutions	Infinitely many Lyapunov stable solutions

Table 4.1. *Stability of the equilibrium states with respect to parameters A and μ*

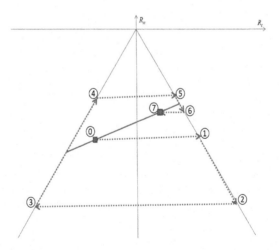

Figure 4.2. *Lyapunov stability of the equilibria in strictly stuck contact for $A > 0$ and $\mu < K_t/W$: the reaction jumps from the initial data to the border of the cone at point 1, then moves smoothy to point 2 where the mass stops out of equilibrium, jumps again to point 3, moves smoothly again to point 4, jumps again to point 5, and moves to a final stop at point 6 where it jumps to the equilibrium 7*

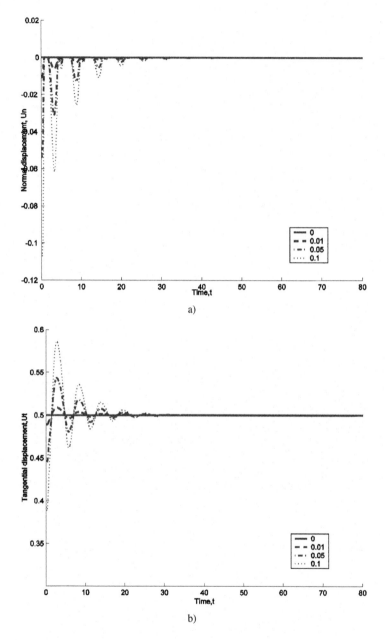

Figure 4.3. *Asymptotic stability of the vertex of the Coulomb's cone for*
$A = 0$ *and* $\mu < K_t/W$

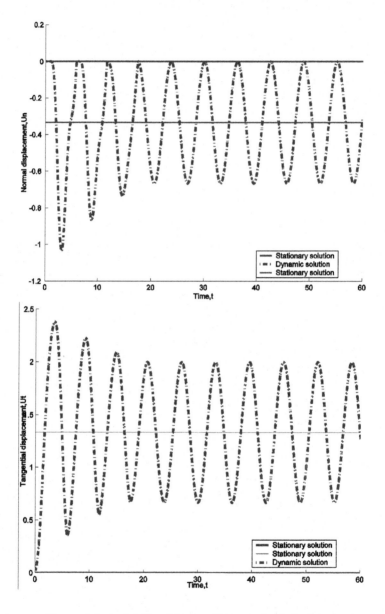

Figure 4.4. *Instability of the equilibrium point in incipient sliding contact for $A < 0$ and $\mu > K_t/W$*

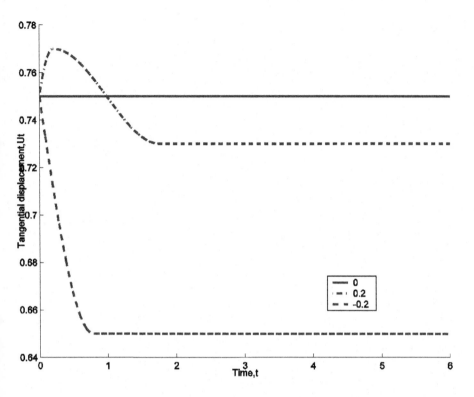

Figure 4.5. *Lyapunov stability of an equilibria in strictly stuck contact for $A > 0$ and $\mu = K_t/W$*

4.2. Introducing a new notion of stability

This section aims at revisiting the basic stability concepts in the case of equilibrium states of discrete systems involving unilateral contact and Coulomb friction. The inequalities induced by the contact and the friction laws, in addition to the dissipative character of Coulomb friction, make it impossible to use the classical stability theorems of discrete systems such as the Lejeune-Dirichlet theorem [APP 04]. Moreover, the graphs of the contact and friction laws rule out any linearization.

Within such a framework, the previous section gave stability properties that were obtained by a direct integration of the dynamics [BAS 03, BAS 06]. If any neighborhood of an equilibrium contains a point that, taken as initial data of the dynamical problem, leads to a trajectory that diverges from the equilibrium, then the equilibrium is said to be unstable. On the contrary, if any point of a neighborhood of the equilibrium taken as initial data leads to a trajectory that tends to the equilibrium or remains in a tubular neighborhood of the equilibrium, then the equilibrium is said to be asymptotically stable, or Lyapunov stable, respectively.

As stressed above, it is clear that such an analysis can be undertaken only after having proved that:

1) the set of equilibria is completely determined for any set of external data (loads, stiffness, friction coefficient, etc.);

2) the Cauchy problem is well-posed (which means that the problem consisting of the equation of the dynamics associated with any admissible initial data has a unique solution);

3) the solution of a discretized problem converges, as the time step tends to zero, toward the solution of the Cauchy problem.

Of course, the fact that the set of equilibria for given forces consists of a single point, a set of isolated points, or a continuous set, indeed even an unbounded set, has an effect on the stability properties. Note that although the existence of a trajectory can hold under very wide conditions, only very smooth data guarantee its uniqueness (see Chapter 3 and [BAL 06]). But in fact:

– The complexity of the problem, simply investigating and classifying the equilibrium states, increases rapidly with the number of degrees of freedom of the system, so that the program consisting of the above three steps – 1), 2) and 3) – has for the moment only been tackled for a mass–spring system containing a single particle;

– The elementary and classical notion of stability that justifies the analysis may not seem totally satisfactory in view of the graph of the Coulomb law. Indeed, an equilibrium solution can be perturbed by a tangential velocity only if it is in imminent sliding. Which means that a given strictly stuck equilibrium solution can be perturbed by a tangential velocity only after the reaction has jumped to the edge of the Coulomb cone, so that even for very small velocities the modification of the reaction may have to be extremely large. This means in turn that it is quite possible that an equilibrium defined by $(U = U^{eq}, \dot{U} = 0)$ is not modified by adding any relatively small external force. Indeed, the deeper are the reactions inside the Coulomb cone the larger the perturbation may have to be.

A new notion of stability, based on the fact that it is equivalent to say that an equilibrium $(U = U^{eq}, \dot{U} = 0)$ is not perturbed by a small enough external force or to say that the corresponding reaction is strictly inside the Coulomb cone, will be stated, and a corresponding conjecture can be qualitatively formulated in the following way:

CONJECTURE 4.1.– Let a discrete system with any finite number of degrees of freedom be at equilibrium under unilateral contact and Coulomb friction conditions. Assume this equilibrium state is such that some reactions are strictly inside the Coulomb cone while the other reactions are in imminent sliding. Then, the trajectory produced by any sufficiently small perturbation of the data leads to a new equilibrium where the number of reactions strictly inside the cone is larger than before the perturbation.

This statement concerns any type of finite dimensional system with unilateral contact and Coulomb friction, which means both granular media, i.e. collections of rigid bodies without any stiffness matrix, and systems having a non-zero stiffness matrix. In this chapter, attention is restricted to mass–spring systems; that is, to cases with a non-zero stiffness matrix. The presence of the stiffness matrix leads to the following corollary:

COROLLARY 4.2.– Let the assumptions be the same as those of conjecture 4.1 in the case of a non-zero stiffness matrix, then if the perturbing force does not depend on time, the final equilibrium is reached in finite time and all the reactions at equilibrium are strictly inside the Coulomb cone.

Having stated the conjecture and its corollary, the remaining part of this chapter aims at backing up these statements by analyzing a few models.

The equations of the dynamics of any discrete system with a linear restoring force can be written in the following abstract form:

$$
\left\{
\begin{array}{l}
\ddot{u} + Ku = F + R, \\
\text{Unilateral contact}, \\
\text{Coulomb friction}, \\
\text{Impact law}, \\
\text{Initial data},
\end{array}
\right.
\qquad [4.8]
$$

so, as used in the previous chapter, the equilibrium equations read:

$$
\left\{
\begin{array}{l}
Ku = F + R, \\
\text{Unilateral contact}, \\
\text{Coulomb friction}.
\end{array}
\right.
\qquad [4.9]
$$

The Coulomb friction law implies that a particle can be set into motion only if its reaction reaches the border of the Coulomb cone. So let an equilibrium state be determined by a pair (u, R) where R is strictly inside the Coulomb cone, then equations [4.9] show that the external forces F can be changed without producing any motion as long as the corresponding reaction R remains strictly inside the cone. This specific property of the equilibrium solutions is due to Coulomb friction. In each of the following models, it is important to give the specific relations that the components of the reaction must satisfy to ensure that this reaction corresponds to an equilibrium solution.

4.3. Analysis of simple models

4.3.1. A chain of masses

The simplest systems involving Coulomb friction are those where the normal component of the reaction is given, such as a mass spring one-dimensional chain moving on a horizontal line, submitted to gravity and to horizontal perturbations in the direction of the line. So the only component of the reaction that varies is the tangential one, and the Coulomb friction law implies that this tangential component must belong to the interval $[-\mu mg, \mu mg]$. If this tangential component is strictly inside the interval $[-\mu mg, \mu mg]$, then the reaction corresponds to an equilibrium solution. This system is a classical model in physics of solids (e.g. [KIT 05]) and has been extensively studied in [PRA 08] in another context. Only simple results relevant to conjecture 4.1 are recalled in the following.

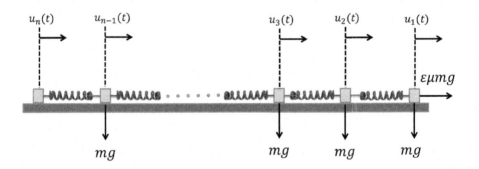

Figure 4.6. *A chain of n masses*

A constant driving force is applied so that the solution is both unique and sufficiently smooth (see [BRE 73]). Figure 4.6 represents the reference configuration. The equations governing motion in the case of two masses linked together by a linearly elastic spring and submitted to Coulomb friction, after a rescaling which involves the mass, the stiffness of the spring and the friction coefficient, are the following:

$$
\begin{cases}
\begin{aligned}
\ddot{u}_1 + (u_1 - u_2) &= r_1 + \varepsilon, \quad t > 0 \\
\ddot{u}_2 + (u_2 - u_1) &= r_2,
\end{aligned} \\[2mm]
-1 \le r_1 \le +1, \quad -1 \le r_2 \le +1, \\[2mm]
\begin{aligned}
u_1(0) &= -1, \; u_2(0) = 0, \\
\dot{u}_1(0) &= \dot{u}_2(0) = 0, \\
r_1(0) &= -1, \; r_2(0) = +1,
\end{aligned}
\end{cases} \tag{4.10}
$$

where r_1 and r_2 are the rescaled tangential components of the reaction (i.e. $r_1 = \dfrac{R_1}{\mu m g}$ and $r_2 = \dfrac{R_2}{\mu m g}$) if R_1 and R_2 are the physical reactions of mass 1 and mass 2, which in this particular case are smooth functions.

The initial values of the tangential component of the reaction correspond to the fact that, before adding the driving force, mass 1 is at rest but in imminent sliding to the right and mass 2 is also at rest but in imminent sliding to the left. This means that mass 1 starts moving as soon as any positive driving force is applied. Note that if the reaction of mass 1 were strictly inside the Coulomb cone, the reaction of mass 2 would also be strictly inside the Coulomb cone (at an equilibrium $r_1 + r_2 = 0$) and therefore any perturbation ε smaller than $\bar{\varepsilon} = r_1(0) + 1$ would fail to put mass 1 into motion. In such a case, the conjecture would be trivially in agreement with experiments.

The motion of mass 1 until it either stops or puts mass 2 into motion is governed by the following differential equation:

$$
\begin{cases}
\ddot{u}_1 + u_1 = -1 + \varepsilon, \quad t > 0, \\
u_1(0) = -1, \dot{u}_1(0) = 0,
\end{cases}
\quad \text{so:} \quad
\begin{cases}
u_1(t) = \varepsilon(1 - \cos t) - 1, \\
\dot{u}_1(t) = \varepsilon \sin t.
\end{cases}
$$

Mass 2 shall not move while $r_2(t) = -u_1(t) = 1 - \varepsilon(1 - \cos t) > -1$ and if $\varepsilon < 1$, then $r_2(t) > -1 \ \forall t$. So, when mass 1 stops the reactions of both masses shall be strictly inside the Coulomb cone, which gives a first case in agreement with the conjecture.

This result holds for a chain of any number of masses as long as the mass next to the one in imminent sliding, on which the external force ε is exerted, is initially in a strictly stuck equilibrium state.

If two adjacent masses are in imminent sliding to the right, the rescaled system which corresponds to the motion of the two masses with external force ε exerted on mass 1 and the reaction $\bar{r}_3$ of mass 3 strictly inside the Coulomb cone, is:

$$\begin{cases} \ddot{u}_1 + (u_1 - u_2) = r_1 + \varepsilon, \\ \ddot{u}_2 + 2u_2 - u_1 - u_3 = r_2, \quad t > 0 \\ u_3 = 0 \\ \\ -1 \leq r_1 \leq +1, \quad -1 \leq r_2 \leq +1, \quad -1 \leq r_3 \leq +1, \\ u_1(0) = -3, \ u_2(0) = -2, \ u_3(0) = 0 \\ \dot{u}_1(0) = \dot{u}_2(0) = 0, \\ r_1(0) = -1, \ r_2(0) = -1, \ r_3(0) = \bar{r}_3 \in\ ]-1, +1[\end{cases} \qquad [4.11]$$

The motion of the two masses modifies the reaction of mass 3 and system [4.11] gives:

$$r_3(t) = \bar{r}_3 - \varepsilon\left(1 + \frac{3 - \sqrt{5}}{2\sqrt{5}} \cos \omega_1 t - \frac{3 + \sqrt{5}}{2\sqrt{5}} \cos \omega_2 t\right),$$

where $\omega_1{}^2 = \dfrac{3 + \sqrt{5}}{2}$ and $\omega_2{}^2 = \dfrac{3 - \sqrt{5}}{2}$.

So, once again for sufficiently small values of the external perturbation ε, reaction $r_3(t)$ stays strictly greater than -1; in other words, the reaction of mass 3 stays strictly inside the Coulomb cone. Therefore, when the first two masses stop moving the reactions of all three masses jump strictly inside the Coulomb cone.

This kind of result, providing in particular a value for a bound on the perturbation ε that leads to an equilibrium solution strictly inside the Coulomb cone, can be easily extended to the case when any number of adjacent masses are simultaneously in imminent sliding on the same side of the cone in the initial configuration. Note that no more than half the number of masses can be concerned as the system is in an equilibrium state before the perturbation is applied (therefore, the sum of all the reactions must be equal to zero).

4.3.2. The single mass model

The model has been presented in Chapter 2, then used in Chapter 3 when dealing with the mathematical foundations of the dynamics, and in Chapter 4 for the complete investigation of the equilibrium states. It is given here as the first example involving a coupling between the normal and the tangential components, although the trajectory is one-dimensional due to the fact that the trajectories remain in contact (see Figure 4.7). The mass shall be in unilateral contact with the horizontal plane and submitted to Coulomb friction. The particle is assumed to be strictly pressed on the horizontal plane so that the normal component of the reaction is strictly negative and sliding motions shall be described by the following differential system:

$$\begin{cases} m\ddot{u}_t + K_t u_t = F_t + R_t, \\ u_n \equiv 0, \qquad\qquad t > 0 \\ W u_t = F_n + R_n, \\[4pt] u_t(0) = u_{t0}, \ \dot{u}_t(0) = 0 \\[4pt] R_n < 0, \\ \mu R_n \le R_t \le -\mu R_n \begin{cases} |R_t| < -\mu R_n \implies \dot{u}_t = 0, \\ |R_t| = -\mu R_n \implies \exists \lambda \ge 0 \ \text{s.t.} \ \dot{u}_t = -\lambda R_t. \end{cases} \end{cases} \qquad [4.12]$$

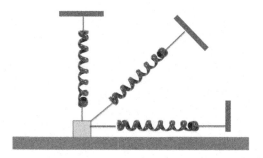

Figure 4.7. *The single mass model*

The equilibrium solutions are all in strict contact when the quantity $A = K_t F_n - W F_t$ is strictly positive. Considering an initial equilibrium where the reactions are strictly inside the Coulomb cone, a perturbation of the force of amplitude $\varepsilon > 0$ can be found such that any perturbation smaller than ε shall leave the mass motionless, so such a case is trivially in agreement with conjecture 4.1. Given (R_t^*, R_n^*) strictly inside the cone, a straightforward calculation gives the radius $\varepsilon > 0$ of a ball centered on (R_t^*, R_n^*) and included in the cone, as observed in Figure 4.8. Therefore, among the equilibrium solutions that are strictly in contact, only those which are in imminent sliding are to be considered. When $K_t - \mu W$ is strictly positive, there are two equilibrium solutions in imminent sliding (one to the right for $R_n = \frac{-K_t F_n + W F_t}{K_t - \mu W}$ and one to the left for $R_n = \frac{-K_t F_n + W F_t}{K_t + \mu W}$), whereas when $K_t - \mu W$ is negative there is only one equilibrium solution in imminent sliding to the left for $R_n = \frac{-K_t F_n + W F_t}{K_t + \mu W}$ (see Chapter 4). The set of normal components of the reaction at time t, corresponding to an equilibrium solution when $K_t - \mu W > 0$, is given by the following segment (corresponding to the projection on the R_n axis of the dotted line, as shown in Figure 4.8):

$$\{R_n\} = \left[\frac{-K_t F_n + W F_t}{K_t - \mu W}, \frac{-K_t F_n + W F_t}{K_t + \mu W} \right],$$

and when $K_t - \mu W \leq 0$ it is a half line given by:

$$\{R_n\} = \left] -\infty, \frac{-K_t F_n + W F_t}{K_t + \mu W} \right].$$

The following lemma characterizes equilibrium solutions.

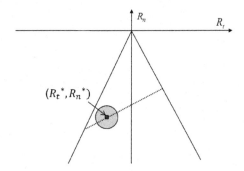

Figure 4.8. *A strictly stuck equilibrium*

LEMMA 4.4.– Let $A > 0$ and the loading F be constant in time. If the trajectory of a sliding mass that satisfies problem [4.12] is such that at the instant t^* when the mass stops sliding its normal reaction R_n^* belongs to the interior of the set $\{R_n\}$ of normal components of the reactions corresponding to equilibrium solutions, then the mass remains in a strictly stuck equilibrium state for all future time.

PROOF.– An existence and uniqueness result for problem [4.12] has been established in [BAL 06] when the loading is piecewise analytical, which is the case when the loading is constant. So that any continuous function whose first derivative is of bounded variation that satisfies all the relations in [4.12] shall be the unique solution of [4.12].

Let $(u_t(t), R_t(t), R_n(t))$ be a solution of [4.12] in $[0, t^*]$ strictly in contact such that $\dot{u}_t(t^*) = 0$ and R_n^* belongs to the interior of the set $\{R_n\}$. Then, this solution can be extended by:

$$\left\{ \begin{array}{l} u_t(t) \equiv u_t(t^*), \quad \dot{u}_t(t) \equiv 0, \\ R_t(t) = K_t u_t(t^*) - F_t, \qquad \forall t > t^* \\ R_n(t) = W u_t(t^*) - F_n. \end{array} \right.$$

$\square$

REMARK 4.1.– Another version of this lemma was given in Chapter 3, where it was needed to establish the computational algorithm.

From now on, the loadings $F_t(t)$ and $F_n(t)$ are chosen in the following way:

$F_t(t) = F_t + P_t(t)$ and $F_n(t) = F_n + P_n(t)$, where $P_t(t)$ and $P_n(t)$ are, respectively, a tangential perturbation and a normal perturbation. For simplicity's sake, only tangential perturbations will be considered. It will be easy to check that adding normal perturbations yields the same results. The equilibrium solution in imminent sliding to the right is considered first. In this case, $K_t - \mu W > 0$ and $u_t = \dfrac{F_t - \mu F_n}{K_t - \mu W}, u_n = 0, R_n = \dfrac{-A}{K_t - \mu W}$ with $R_t = \mu R_n$.

A constant tangential perturbation ε is then applied. In this case, the set of normal components of the reaction corresponding to equilibrium solutions is time independent and given by:

$$\{R_n^\varepsilon\} = \left[\frac{-A + \varepsilon W}{K_t - \mu W}, \frac{-A + \varepsilon W}{K_t + \mu W} \right].$$

If the perturbation ε is strictly negative, then the reaction jumps to a value strictly inside the Coulomb cone and the mass is in a strictly stuck equilibrium state. On the other hand, if ε is strictly positive, then the mass starts sliding to the right and its motion satisfies the following differential equation:

$$\left\{ \begin{array}{l} m\ddot{u}_t + (K_t - \mu W)u_t = F_t + \varepsilon - \mu F_n, \\ u_t(0) = \dfrac{F_t - \mu F_n}{K_t - \mu W}, \quad \dot{u}_t(0) = 0, \end{array} \right. \qquad t > 0. \qquad [4.13]$$

The solution of this equation is given by:

$$u_t(t) = \frac{F_t - \mu F_n}{K_t - \mu W} + \frac{\varepsilon}{K_t - \mu W}(1 - \cos \omega_\alpha t),$$

where ω_α is the intrinsic frequency of the sliding to the right, introduced at the beginning of this chapter, given by $\omega_\alpha^2 = \dfrac{K_t - \mu W}{m}$. When the mass stops sliding at $t^* = \dfrac{\pi}{\omega_\alpha}$, the tangential component of the displacement is given by:

$$u_t(t^*) = \frac{F_t - \mu F_n}{K_t - \mu W} + \frac{2\varepsilon}{K_t - \mu W},$$

and the normal component of the reaction by:

$$R_n^\star = \frac{-A + 2\varepsilon W}{K_t - \mu W}.$$

It is immediately seen that if $\varepsilon < \dfrac{2\mu A}{K_t + 3\mu W}$, then

$$R_n^\star \in \left] \frac{-A + \varepsilon W}{K_t - \mu W}, \frac{-A + \varepsilon W}{K_t + \mu W} \right[.$$

In other words, when the mass stops sliding there is a jump in the tangential component of the reaction, and for sufficiently small values of the perturbation, the trajectory leads to an equilibrium state strictly inside the Coulomb cone.

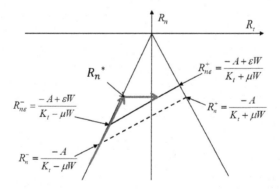

Figure 4.9. *A perturbation of the set of equilibria (dashed line: non-perturbed equilibrium solutions; dotted line: perturbed equilibrium solutions)*

A similar computation establishes that if the equilibrium state that is in imminent sliding to the left is considered, any positive tangential perturbation shall make the tangential reaction jump strictly inside the Coulomb cone, so that the mass shall be in a strictly stuck equilibrium state. A negative tangential perturbation shall make the mass move but when it stops sliding, the jump in the tangential component of the reaction shall bring the reaction strictly inside the Coulomb cone if $|\varepsilon|(K_t - 3\mu W) < 2\mu A$ when $K_t - \mu W > 0$ and for all values of ε when $K_t - \mu W \leq 0$.

4.4. A slightly more complicated mass–spring system

The problem considered here is shown in Figure 4.10. This system presented in Chapter 2 is simply built by coupling two single mass–spring models of the type shown in Figure 4.7. It has first been studied in [ALA 86] and more recently in [PIN 01]. The two particles are of mass m, the stiffness of the springs is equal to k and φ is the angle between the springs. In the following, the notations $c := \cos\varphi$ and $s := \sin\varphi$ have been introduced. The movement of the two masses is governed by [4.14]–[4.18] where the parameters k and m are chosen equal to 1 in equations [4.14] (an adequate rescaling would have had the same effect):

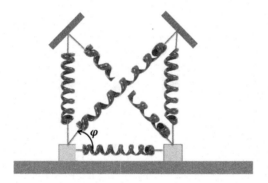

Figure 4.10. *The two-mass problem*

– *The equations*:

$$\begin{cases} \ddot{u}_{1t} + (1 + c^2)u_{1t} + csu_{1n} - u_{2t} = F_{1t} + R_{1t}, \\ \ddot{u}_{1n} + csu_{1t} + (1 + s^2)u_{1n} = F_{1n} + R_{1n}, \\ \ddot{u}_{2t} - u_{1t} + (1 + c^2)u_{2t} - csu_{2n} = F_{2t} + R_{2t}, \\ \ddot{u}_{2n} - csu_{2t} + (1 + s^2)u_{2n} = F_{2n} + R_{2n}. \end{cases} \qquad [4.14]$$

– *The initial conditions*:

For $i = 1, 2$ $u_{it}(0) = u_{it0}, \dot{u}_{it}(0) = v_{it0}, u_{in}(0) = u_{in0}, \dot{u}_{in}(0) = v_{in0}.$ [4.15]

– *The unilateral contact conditions*:

For $i = 1, 2$ $R_{in} \leq 0,$ $u_{in} \leq 0,$ $R_{in}u_{in} = 0.$ [4.16]

– *The Coulomb friction law*:

For $i = 1, 2$

$R_{in} = 0$ then $\dot{u}_{it} \in] - \infty, +\infty[$

$|R_{it}| \leq -\mu R_{in}$ and $\begin{cases} |R_{it}| < -\mu R_{in} \implies \dot{u}_{it} = 0, \\ |R_{it}| = -\mu R_{in} \implies \exists \lambda \geq 0 \text{ s.t. } \dot{u}_{it} = -\lambda R_{it}. \end{cases}$ [4.17]

– *The impact law*:

For $i = 1, 2$ when $u_{in}(t) = 0,$
$\dot{u}_{in}(t^+) = -e\dot{u}_{in}(t^-)$ with e $\in [0, 1].$ [4.18]

In system [4.14]–[4.18], R_{it} and R_{in}, $i = 1, 2$, are, respectively, the tangential and the normal components of the reaction exerted by the obstacle on mass 1 and mass 2, μ is the friction coefficient, u_{it} and u_{in} the tangential and the normal components of the displacement and ($\dot{}$) stands for the time derivative. F_{it} and F_{in} are the tangential and normal components of the external loading on each mass.

4.4.1. Equilibrium states

The equilibrium solutions of problem [4.14–4.18] are determined, in the case where the external forces are constant, by solving the following system, generalizing the method presented in Chapter 4:

$$\begin{cases} (1 + c^2)u_{1t} + csu_{1n} - u_{2t} = F_{1t} + R_{1t}, \\ csu_{1t} + (1 + s^2)u_{1n} = F_{1n} + R_{1n}, \\ -u_{1t} + (1 + c^2)u_{2t} - csu_{2n} = F_{2t} + R_{2t}, \\ -csu_{2t} + (1 + s^2)u_{2n} = F_{2n} + R_{2n}, \end{cases}$$ [4.19]

For $i = 1, 2$ $\begin{cases} R_{in} \leq 0, \quad u_{in} \leq 0, \quad R_{in}u_{in} = 0, \\ |R_{it}| \leq -\mu R_{in}. \end{cases}$ [4.20]

An equilibrium solution is given by the set $(u_{in}, u_{it}, R_{in}, R_{it})$ for $i = 1, 2$ that satisfies both [4.19] and [4.20]. Due to conditions [4.20], these relations define a strongly nonlinear system; however, looking for a solution with no contact requires inserting $R_{in} = 0$, $i = 1, 2$ into system [4.19] and the displacements must satisfy [4.20], whereas looking for a solution in contact requires inserting $u_{in} = 0$, $i = 1, 2$ into system [4.19]; in this case the reactions must satisfy [4.20]. When dealing with the one mass case, it was relatively easy to describe extensively the different equilibrium states according to the different values of the parameters and these equilibrium states could be summarized in a table. Here such a table would be much too complicated, so attention will be restricted to simply describing the different equilibrium states and giving the values of the parameters that lead to such states.

4.4.1.1. The two masses are not in contact

The displacements of the equilibrium solution corresponding to the no contact case are obtained by solving the linear system [4.19] with $R_{in} = 0$, $i = 1, 2$. However, this shall correspond to an equilibrium solution only if conditions [4.20] are fulfilled. As the reactions are zero, these conditions reduce to $u_{in} \leq 0$, $i = 1, 2$. The normal components of the displacements are given by:

$$u_{1n} = \frac{\breve{A}_1}{4c^2 - c^4} \text{ and } u_{2n} = \frac{\breve{A}_2}{4c^2 - c^4},$$

where quantities $\breve{A}_1$ and $\breve{A}_2$ depend only on the data and are defined by:

$$\begin{aligned} \breve{A}_1 &= 3c^2 F_{1n} - s^2 c^2 F_{2n} - 2sc F_{1t} - sc(1 + s^2) F_{2t}, \\ \breve{A}_2 &= 3c^2 F_{2n} - s^2 c^2 F_{1n} + 2sc F_{2t} + sc(1 + s^2) F_{1t}. \end{aligned} \qquad [4.21]$$

As $4c^2 - c^4$ is strictly positive, the solution of [4.19] thus obtained is an equilibrium solution only if $\breve{A}_1 \leq 0$ and $\breve{A}_2 \leq 0$. If either $\breve{A}_1 > 0$ or $\breve{A}_2 > 0$, there is no equilibrium solution where the two masses are not in contact.

4.4.1.2. Only one mass is in contact

Mass 2 is assumed to be in contact and mass 1 not. The problem to be solved to determine the set of equilibrium solutions is then:

$$\begin{cases} (1 + c^2) u_{1t} + cs u_{1n} - u_{2t} = F_{1t}, \\ cs u_{1t} + (1 + s^2) u_{1n} = F_{1n}, \\ -u_{1t} + (1 + c^2) u_{2t} = F_{2t} + R_{2t}, \\ -cs u_{2t} = F_{2n} + R_{2n}, \end{cases} \qquad [4.22]$$

together with

$$u_{1n} \leq 0, \quad u_{2n} = 0, \quad R_{1t} = R_{1n} = 0, \quad R_{2n} \leq 0, \quad |R_{2t}| \leq -\mu R_{2n}. \qquad [4.23]$$

By expressing u_{1t} and u_{2t} in terms of the data and of R_{2t} and R_{2n}, then eliminating u_{1n}, the following relationship between the normal and tangential components of the reaction of mass 2 is obtained:

$$F_{1t} - \frac{F_{2n} + R_{2n}}{cs} + (1 + c^2)\left[F_{2t} + R_{2t} + \frac{1 + c^2}{cs}(F_{2n} + R_{2n})\right]$$
$$= \frac{cs}{1 + s^2}\left[F_{1n} + cs(F_{2t} + R_{2t}) + (1 + c^2)(F_{2n} + R_{2n})\right].$$ [4.24]

Introducing into the above expression $\breve{A}_2$ (defined in [4.21]) leads to:

$$R_{2n} = -\frac{2s}{3c}R_{2t} - \frac{\breve{A}_2}{3c^2}.$$

– if $\breve{A}_2 < 0$ and $\mu \leq \dfrac{3c}{2s}$, no equilibrium solution exists because no pair (R_{2t}, R_{2n}) satisfies both the above relation and condition [4.23];

– if $\breve{A}_2 < 0$ and $\mu > \dfrac{3c}{2s}$, then any pair (R_{2t}, R_{2n}) satisfying:

$$R_{2n} \leq \frac{\breve{A}_2}{c(2s\mu - 3c)} \quad \text{and} \quad R_{2t} = -\frac{3c}{2s}R_{2n} - \frac{\breve{A}_2}{3cs}$$

corresponds to an equilibrium solution as long as $u_{1n} \leq 0$. The normal displacement of mass 1, u_{1n}, is then given by:

$$u_{1n} = \frac{1}{1 + s^2}\left[(1 - \frac{c^2}{2})R_{2n} - \frac{2}{2} + F_{1n} + csF_{2t} + (1 + c^2)F_{2n}\right].$$

At this point, it is useful to introduce two new quantities that depend only on the data:

$$d_1 \overset{def}{\equiv} (1 + c^2)F_{1n} + F_{2n} - csF_{1t},$$
$$d_2 \overset{def}{\equiv} F_{1n} + (1 + c^2)F_{2n} + csF_{2t}.$$ [4.25]

It is easily seen that quantities d_1 and d_2 are related to $\breve{A}_1$ and $\breve{A}_2$ in the following way:

$$\breve{A}_1 = 2d_1 + (c^2 - 2)d_2$$
$$\breve{A}_2 = (c^2 - 2)d_1 + 2d_2$$ [4.26]

So that finally the normal component of the displacement of mass 1 can be expressed as:

$$u_{1n} = \frac{R_{2n} + d_1}{2}.$$ [4.27]

If $d_1 \leq 0$, then by equation [4.27] all the pairs (R_{2t}, R_{2n}) defined above are compatible with the unilateral conditions, whereas if $d_1 > 0$ only the pairs such that $R_{2n} \leq -d_1$ give rise to an equilibrium solution. Figures 4.11(a) and (b) represent in the $\{R_{2t}, R_{2n}\}$ plane the sets of (R_{2t}, R_{2n}) corresponding to an equilibrium solution when $d_1 \leq 0$ and when $d_1 > 0$, respectively. This figure is the same as in the one mass case when $d_1 \leq 0$, but when $d_1 > 0$ the pairs of (R_{2t}, R_{2n}) corresponding to an equilibrium solution must also satisfy $R_{2n} \leq -d_1$ because of [4.23] and [4.27]. The results obtained, when mass 2 is not in contact and mass 1 is, are given by similar expressions where the subscripts 1 and 2 only have to be interchanged.

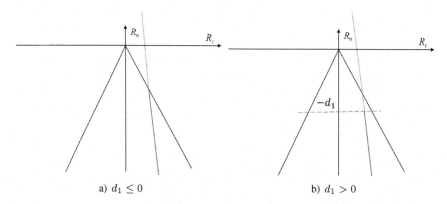

a) $d_1 \leq 0$ b) $d_1 > 0$

Figure 4.11. *Equilibrium reactions in the* $\{R_{2t}, R_{2n}\}$ *plane*

4.4.1.3. *The two masses are in contact*

When the two masses are in contact both u_{1n} and u_{2n} are equal to zero, so the equilibrium solution is obtained by solving the following problem:

$$\begin{cases} (1 + c^2)u_{1t} - u_{2t} = F_{1t} + R_{1t}, \\ csu_{1t} = F_{1n} + R_{1n}, \\ -u_{1t} + (1 + c^2)u_{2t} = F_{2t} + R_{2t}, \\ -csu_{2t} = F_{2n} + R_{2n}, \end{cases}$$ [4.28]

with:

$$R_{1n} \leq 0, \quad R_{2n} \leq 0, \quad u_{1n} = 0, \quad u_{2n} = 0,$$
$$|R_{1t}| \leq \mu R_{1n}, \quad |R_{2t}| \leq \mu R_{2n}. \tag{4.29}$$

Eliminating u_{1t} and u_{2t} gives:

$$(1 + c^2)(F_{1n} + R_{1n}) + F_{2n} + R_{2n} = cs(F_{1t} + R_{1t}),$$
$$-(F_{1n} + R_{1n}) - (1 + c^2)(F_{2n} + R_{2n}) = cs(F_{2t} + R_{2t}), \tag{4.30}$$

which can be rewritten as:

$$(1 + c^2)R_{1n} + R_{2n} - csR_{1t} = -\left[(1 + c^2)F_{1n} + F_{2n} - csF_{1t}\right]$$
$$R_{1n} + (1 + c^2)R_{2n} + csR_{2t} = -\left[F_{1n} + (1 + c^2)F_{2n} + csF_{2t}\right]. \tag{4.31}$$

Introducing d_1 and d_2 defined by [4.25] into the above expressions, the following relations between R_{1n}, R_{2n}, R_{1t} and R_{2t} are obtained:

$$\begin{cases} (1 + c^2)R_{1n} + R_{2n} - csR_{1t} = -d_1 \\ R_{1n} + (1 + c^2)R_{2n} + csR_{2t} = -d_2. \end{cases} \tag{4.32}$$

Vector $u = (R_{1t}, R_{1n}, R_{2t}, R_{2n})$ belongs to $\mathbb{R}^4$ so that the vectors u satisfying the above relations determine a plane $(\mathcal{P})$ in $\mathbb{R}^4$ defined by:

$$u \in (\mathcal{P}) \iff u = \frac{1}{cs}\begin{pmatrix} d_1 \\ 0 \\ -d_2 \\ 0 \end{pmatrix} + \frac{\alpha}{cs}\begin{pmatrix} 1 + c^2 \\ cs \\ -1 \\ 0 \end{pmatrix} + \frac{\beta}{cs}\begin{pmatrix} 1 \\ 0 \\ -(1 + c^2) \\ cs \end{pmatrix}$$

with $(\alpha, \beta) \in \mathbb{R}^2$.

Relations [4.29] can be expressed in the $\{\alpha, \beta\}$ plane in the following way:

$$R_{1n} \leq 0 \iff \alpha \leq 0,$$
$$R_{2n} \leq 0 \iff \beta \leq 0.$$

$$R_{1t} \leq -\mu R_{1n} \iff d_1 + \alpha[(1+c^2) + \mu cs] + \beta \leq 0,$$
$$R_{1t} \geq \mu R_{1n} \iff d_1 + \alpha[(1+c^2) - \mu cs] + \beta \geq 0,$$
$$R_{2t} \leq -\mu R_{2n} \iff d_2 + \beta[(1+c^2) - \mu cs] + \alpha \geq 0,$$
$$R_{2t} \geq \mu R_{2n} \iff d_2 + \beta[(1+c^2) + \mu cs] + \alpha \leq 0.$$ [4.33]

Relations [4.33] suggest the introduction of two new constants $\mathcal{C}$ and $\mathcal{C}'$ defined as:

$$\mathcal{C} \overset{def}{\equiv} (1+c^2) + \mu cs, \quad \mathcal{C}' \overset{def}{\equiv} (1+c^2) - \mu cs.$$ [4.34]

Relations [4.33] define a domain of the $\{\alpha, \beta\}$ plane, and therefore of the $\{R_{1n}, R_{2n}\}$ plane. It is easy to see that if $\mu < (1+c^2)/cs$ and either d_1 or d_2 are negative, then the domain defined by [4.33] is empty. Figures 4.12 and 4.13 show different situations that can occur when both d_1 and d_2 are strictly positive according to different values of μ. The values of R_{1n} and R_{2n} on the boundary of the domain correspond to equilibrium solutions that are in imminent sliding, and the shape of the $\{R_{1n}, R_{2n}\}$ equilibria domain depends on the parameters and may be unbounded.

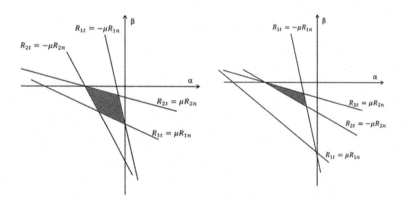

Figure 4.12. *Bounded domains of equilibrium in the* $\{\alpha(= R_{1n}), \beta(= R_{2n})\}$ *plane*

The results of this section show that for any given sign of the data $\breve{A}_1, \breve{A}_2, d_1, d_2$, and according to whether μ is smaller than $3c/2s$, greater than $(1+c^2)/cs$ or between both values, all the possible equilibrium solutions can be explicitly determined. In general, equilibrium solutions where only one mass is in contact and solutions where both masses are in contact coexist. However, when $\breve{A}_1 < 0$, $\breve{A}_2 < 0$ and $\mu < 3c/2s$, the only possible equilibrium solution is the solution where the two masses are not in contact. When $\breve{A}_1 = 0$, $\breve{A}_2 = 0$ and $\mu < 3c/2s$, the only equilibrium solution is the one where the two masses are in grazing contact.

REMARK 4.2.– It is interesting to note the complexity of the results obtained here compared to those obtained in the case of a single mass. In the case of a single mass, the nine different sets of equilibrium solutions according to the values of the data were represented (see Chapter 4, section 4.1.2). In the case of two masses there would be more than 30 different sets of equilibrium solutions, and inside a given set we can have infinitely many solutions where the two masses are in contact together, and still infinitely many solutions where only one mass is in contact. There is no point in giving all these different sets, but through the analysis given in this section the set corresponding to a given data can easily be determined.

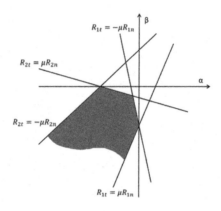

Figure 4.13. *Unbounded domains of equilibrium in the* $\{\alpha(= R_{1n}), \beta(= R_{2n})\}$ *plane*

4.4.2. *Stability*

Having obtained a complete description of the equilibrium solutions, it is now possible to proceed in the endeavor to support conjecture 4.1.

In the case of two masses, the conjecture concerns a perturbation of an equilibrium where the two masses are strictly in contact. If one of the two masses is not in contact, the motion of the two masses obtained by perturbing the mass strictly in contact can be explicitly computed, and we can check that after a sliding of the mass in contact (or eventually several right and left slidings) the mass in contact shall stay motionless, whereas the mass which is not in contact shall oscillate indefinitely. So, a new equilibrium state is never reached. If both reactions are strictly inside the Coulomb cone, then there exist small enough perturbations such that the reactions stay strictly inside the Coulomb cone, in which case the conjecture is trivial. Consequently, the result remains to be established only in the case where at least one mass is in imminent sliding. The case where only one mass is set into motion by the perturbation is studied first, before dealing with the case where both masses are set into motion. When dealing with the two mass case, a perturbation is added to the normal component, whereas in the case of a single mass, it is the tangential component being perturbed. Indeed, in the case of a single mass, a normal perturbation would lead to trivial results but in the case of two masses, even a normal perturbation yields interesting results.

4.4.2.1. *One mass is stuck, the other is in imminent sliding*

Problems [4.14]–[4.18] are considered with initial data corresponding to an equilibrium where only one mass is in imminent sliding. A constant perturbation ε of the force is applied, and it is now shown that the trajectory leads to a new equilibrium where both reactions are strictly inside the cone. Assume the initial state is such that the reaction of mass 1 is strictly inside the Coulomb cone and that mass 2 is in imminent sliding to the right (for example when the parameters correspond to the reactions shown in Figure 4.12). The initial equilibrium solution then satisfies $u_{1n} = u_{2n} = 0$, $u_{1t} = \bar{u}_{1t}$, $u_{2t} = \bar{u}_{2t}$, $R_{1n} = R_{1n}^{eq}$, $R_{2n} = R_{2n}^{eq}$, $|R_{1t}^{eq}| < -\mu R_{1n}^{eq}$ and $R_{2t}^{eq} = \mu R_{2n}^{eq}$, where:

$$
\bar{u}_{1t} = \frac{F_{1n} + R_{1n}^{eq}}{sc}, \quad \bar{u}_{2t} = -\frac{F_{2n} + R_{2n}^{eq}}{sc} := \frac{\mathcal{D}}{\mathcal{C}},
$$

with $\mathcal{C}$ defined above in [4.34] and $\mathcal{D}$ given by:

$$
\mathcal{D} = F_{2t} - \mu F_{2n} + \frac{F_{1n} + R_{1n}^{eq}}{sc}.
$$

For such an equilibrium solution, both R_{1n}^{eq} and R_{2n}^{eq} are strictly negative. Assume the perturbation ε is sufficiently small to ensure that the normal reactions of both masses remain strictly negative and that the reaction of mass 1 stays strictly inside the Coulomb cone. Since the equilibrium solution is such that the reaction of mass 1 is strictly inside the Coulomb cone and mass 2 is in imminent sliding, equations [4.33] and [4.34] imply that R_{1n}^{eq} and R_{2n}^{eq} must satisfy the following relations:

$$
\begin{cases}
d_1 + \mathcal{C} R_{1n}^{eq} + R_{2n}^{eq} < 0, \\
d_1 + \mathcal{C}' R_{1n}^{eq} + R_{2n}^{eq} > 0, \\
d_2 + \mathcal{C} R_{2n}^{eq} + R_{1n}^{eq} = 0, \\
d_2 + \mathcal{C}' R_{2n}^{eq} + R_{1n}^{eq} > 0.
\end{cases}
\qquad [4.35]
$$

If some positive constant normal force ε is added to the normal component F_{2n} of the loading applied on mass 2, then the reaction of mass 2 enters strictly inside the

Coulomb cone and mass 2 stays motionless. But if the perturbation ε applied to F_{2n} is negative then problem [4.14]–[4.18] becomes:

$$\begin{cases} u_{1t}(t) = \bar{u}_{1t}, \ \ u_{1n}(t) = 0, \ \ u_{2n}(t) = 0 \ \forall t, \\ \\ \ddot{u}_{2t} + Cu_{2t} = F_{2t} - \mu(F_{2n} + \varepsilon) + \dfrac{F_{1n} + R_{1n}^{eq}}{sc} = \mathcal{D} - \mu\varepsilon, \\ u_{2t}(0) = \bar{u}_{2t} = \dfrac{\mathcal{D}}{C}, \\ \dot{u}_{2t}(0) = 0, \end{cases} \qquad [4.36]$$

and therefore

$$u_{2t}(t) = \frac{\mathcal{D} - \mu\varepsilon}{C} + \frac{\mu\varepsilon}{C} \cos \sqrt{C}t. \qquad [4.37]$$

The state when mass 2 stops sliding will be the final state of the system, with both reactions strictly inside the Coulomb cone. This is proved as follows.

Let $\tilde{t} > 0$ be such that $\dot{u}_{2t}(\tilde{t}) = 0$ (i.e. $\tilde{t}$ is the instant when mass 2 stops sliding), then $u_{2t}(\tilde{t}) = \dfrac{\mathcal{D} - 2\mu\varepsilon}{C}$. As mass 2 always remains in contact, Coulomb's law and the equations of motion imply that there is no velocity jump at $t = \tilde{t}$ (see [BAL 06]). So $\dot{u}_{2t}^{+}(\tilde{t}) = 0$. It shall be shown that η exists such that $\dot{u}_{2t}(t) = 0 \ \forall t \in]\tilde{t}, \tilde{t} + \eta[$; in other words, that mass 2 stays motionless after $\tilde{t}$.

If $\dot{u}_{2t}(t) > 0 \ \forall t \in]\tilde{t}, \tilde{t} + \eta[$, then mass 2 continues to slide to the right (i.e. $R_{2t} = \mu R_{2n}$). Consequently, the motion is still described by system [4.36] in $]\tilde{t}, \tilde{t} + \eta[$ and there is no jump of the tangential component of the acceleration of mass 2. Therefore, $\ddot{u}_{2t}(\tilde{t}) = \mu\varepsilon < 0$ and thus $\dot{u}_{2t}(t) < 0$ for $t \in]\tilde{t}, \tilde{t} + \eta[$, which contradicts the assumption.

Conversely, if $\dot{u}_{2t}(t) < 0$ for $t \in]\tilde{t}, \tilde{t} + \eta[$, then mass 2 will slide to the left (i.e. $R_{2t} = -\mu R_{2n}$) and u_{2t} satisfies the following system for $t \in]\tilde{t}, \tilde{t} + \eta[$:

$$\begin{cases} \ddot{u}_{2t} + C'u_{2t} = \mathcal{D}' + \mu\varepsilon, \\ u_{2t}(\tilde{t}) = \dfrac{\mathcal{D} - 2\mu\varepsilon}{C}, \dot{u}_{2t}(\tilde{t}) = 0, \end{cases} \qquad [4.38]$$

where $\mathcal{D}' = F_{2t} + \mu F_{2n} + \bar{u}_{1t}$, so that for $t \in]\tilde{t}, \tilde{t} + \eta[$,

$$\ddot{u}_{2t}(t) = -\frac{1}{C}(C'(\mathcal{D} - 2\mu\varepsilon) - C(\mathcal{D}' + \mu\varepsilon)).$$

But as $C'(\mathcal{D} - 2\mu\varepsilon) - C(\mathcal{D}' + \mu\varepsilon) = 2\mu C R_{2n}^{eq} - (3(1 + c^2) - \mu cs)\mu\varepsilon$, $\varepsilon_0 > 0$ exists such that if $|\varepsilon| < \varepsilon_0$, this quantity is strictly negative and $\ddot{u}_{2t}(t) > 0$. So that $\dot{u}_{2t}$ is positive in some interval $]\tilde{t}, \tilde{t} + \eta[$, which also contradicts the assumption.

Therefore, for $t \in]\tilde{t}, \tilde{t} + \eta[$, the sliding velocity $\dot{u}_{2t}(t)$ of mass 2 remains zero, so the displacements and reactions are solution to system [4.19] and [4.20] in this time interval, which implies that at time $\tilde{t}$ there is a jump of the tangential reaction of mass 2 and of its tangential acceleration, after which

$$\ddot{u}_{2t}(t) = 0, \quad \text{for } t \in]\tilde{t}, \tilde{t} + \eta[\text{ and } \tilde{R}_{2t}^+ = R_{2t}^{eq} - 2\mu\frac{1 + c^2}{C}\varepsilon.$$

Then, the normal component of the reaction of mass 2 at $\tilde{t}$ is

$$\tilde{R}_{2n} = R_{2n}(\tilde{t}) = -csu_{2t}(\tilde{t}) - F_{2n} - \varepsilon = R_{2n}^{eq} - \varepsilon\frac{C'}{C}, \qquad [4.39]$$

and a simple computation ensures that there exists $\varepsilon_0 > 0$ such that if $|\varepsilon| < \varepsilon_0$, then R_{1n}^{eq} and $\tilde{R}_{2n}$ satisfy the following relations:

$$\begin{cases} \tilde{d}_1 + C R_{1n}^{eq} + \tilde{R}_{2n} < 0, \\ \tilde{d}_1 + C' R_{1n}^{eq} + \tilde{R}_{2n} > 0, \\ \tilde{d}_2 + C \tilde{R}_{2n} + R_{1n}^{eq} < 0, \\ \tilde{d}_2 + C' \tilde{R}_{2n} + R_{1n}^{eq} > 0, \end{cases} \qquad [4.40]$$

where $\tilde{d}_1 = d_1 + \varepsilon$ and $\tilde{d}_2 = d_2 + (1 + c^2)\varepsilon$.

This proves the conjecture when mass 2 is in imminent sliding to the right. When mass 2 is in imminent sliding to the left and F_{2n} is perturbed by adding $\varepsilon > 0$, then the reaction of mass 2 is strictly inside the Coulomb cone and mass 2 stays motionless, while if $\varepsilon < 0$ is added to F_{2n} then mass 2 starts to slide to the left. A similar computation to the one given above establishes that there exists $\varepsilon_0 > 0$ such that if $|\varepsilon| < \varepsilon_0$, then when mass 2 stops sliding to the left its reaction is strictly inside the Coulomb cone and therefore that mass 2 stays motionless.

4.4.2.2. Perturbing both masses in imminent sliding

When both masses are in imminent sliding, one to the left and the other to the right, applying a positive perturbation to the normal loading of mass 2 leaves the reaction of mass 2 strictly inside the Coulomb cone and mass 1 in imminent sliding, whereas a negative perturbation initiates the sliding of mass 2, and this sliding makes the reaction of mass 1 lie strictly inside the Coulomb cone. In both cases, applying a perturbation to any one mass leaves the two masses in the situation studied in the previous section.

Therefore, a perturbation to the normal component of the loading is applied to both masses which are both in imminent sliding. Let mass 1 be in imminent left sliding and mass 2 in imminent right sliding, that is:

$$d_1 + CR^{eq}_{1n} + R^{eq}_{2n} = 0 \text{ and } d_2 + CR^{eq}_{2n} + R^{eq}_{1n} = 0.$$

Adding ε_1 to the loading F_{1n} and ε_2 to the loading F_{2n} changes the parameters d_1 and d_2, which become:

$$\tilde{d}_1 = d_1 + (1 + c^2)\varepsilon_1 + \varepsilon_2 \text{ and } \tilde{d}_2 = d_2 + \varepsilon_1 + (1 + c^2)\varepsilon_2.$$

So the normal reactions become after the perturbation:

$$R_{1n}(0) = R^{eq}_{1n} - \varepsilon_1 \text{ and } R_{2n}(0) = R^{eq}_{2n} - \varepsilon_2.$$

Therefore

$$\tilde{d}_1 + CR_{1n}(0) + R_{2n}(0) = -\mu cs\varepsilon_1,$$

$$\tilde{d}_2 + CR_{2n}(0) + R_{1n}(0) = -\mu cs\varepsilon_2.$$

From these expressions, it is immediately seen that if ε_1 is positive the reaction of mass 1 is in the Coulomb cone and if ε_2 is positive the reaction of mass 2 is in the Coulomb cone. As the case when only one perturbation is negative has been studied in the previous section, suppose now that both ε_1 and ε_2 are negative.

The motion of the two masses satisfies the following system:

$$\begin{cases} \ddot{u}_{1t} + Cu_{1t} - u_{2t} = F_{1t} + \mu F_{1n} + \mu\varepsilon_1, \\ \ddot{u}_{2t} - u_{1t} + Cu_{2t} = F_{2t} - \mu F_{2n} - \mu\varepsilon_2, \\ \\ u_{1t}(0) = \bar{u}_{1t} = \dfrac{C(F_{1t} + \mu F_{1n}) + F_{2t} - \mu F_{2n}}{C^2 - 1}, \quad \dot{u}_{1t}(0) = 0, \\ u_{2t}(0) = \bar{u}_{2t} = \dfrac{C(F_{2t} - \mu F_{2n}) + F_{1t} + \mu F_{1n}}{C^2 - 1}, \quad \dot{u}_{2t}(0) = 0, \end{cases} \qquad [4.41]$$

which gives

$$u_{1t}(t) = \bar{u}_{1t} + \mu\frac{C\varepsilon_1 - \varepsilon_2}{C^2 - 1} + \mu\frac{\varepsilon_2 - \varepsilon_1}{2\omega_1^2}\cos\omega_1 t - \mu\frac{\varepsilon_1 + \varepsilon_2}{2\omega_2^2}\cos\omega_2 t,$$

[4.42]

$$u_{2t}(t) = \bar{u}_{2t} + \mu\frac{\varepsilon_1 - C\varepsilon_2}{C^2 - 1} + \mu\frac{\varepsilon_2 - \varepsilon_1}{2\omega_1^2}\cos\omega_1 t + \mu\frac{\varepsilon_1 + \varepsilon_2}{2\omega_2^2}\cos\omega_2 t.$$

$$\dot{u}_{1t}(t) = -\frac{\mu}{2\omega_1}(\varepsilon_2 - \varepsilon_1)\sin\omega_1 t + \frac{\mu}{2\omega_2}(\varepsilon_1 + \varepsilon_2)\sin\omega_2 t,$$

[4.43]

$$\dot{u}_{2t}(t) = -\frac{\mu}{2\omega_1}(\varepsilon_2 - \varepsilon_1)\sin\omega_1 t - \frac{\mu}{2\omega_2}(\varepsilon_1 + \varepsilon_2)\sin\omega_2 t,$$

where $\omega_1 = \sqrt{C - 1}$ and $\omega_2 = \sqrt{C + 1}$.

If ε_1 is different from ε_2, let $\varepsilon_1 < \varepsilon_2 < 0$. In this case, mass 2 stops before mass 1. Let $\tilde{t}$ be such that $\dot{u}_{2t}(\tilde{t}) = 0$ and $\dot{u}_{2t}(t) > 0$ for $0 < t < \tilde{t}$. It shall now be established that the further dynamics is such that $\dot{u}_{2t}(t) = 0\ \forall t > \tilde{t}$.

If η exists such that $\dot{u}_{2t}(t) > 0\ \forall t \in]\tilde{t}, \tilde{t} + \eta[$, then mass 2 continues to slide to the right and the motion continues to satisfy system [4.41] so that the corresponding solution is sufficiently smooth in $]\tilde{t} - \eta, \tilde{t} + \eta[$ for the third derivative to exist:

$$\dddot{u}_{2t}(\tilde{t}) = \frac{\mu}{2}\left(\omega_1(\varepsilon_2 - \varepsilon_1)\sin\omega_1\tilde{t} + \omega_2(\varepsilon_1 + \varepsilon_2)\sin\omega_2\tilde{t}\right)$$

but $\dot{u}_{2t}(\tilde{t}) = 0 \implies \dfrac{\varepsilon_1 + \varepsilon_2}{\omega_2}\sin\omega_2\tilde{t} = -\dfrac{\varepsilon_2 - \varepsilon_1}{\omega_1}\sin\omega_1\tilde{t}$

then

$$\dddot{u}_{2t}(\tilde{t}) = -\mu\frac{\varepsilon_2 - \varepsilon_1}{\omega_1}\sin\omega_1\tilde{t}.$$

Inserting $\bar{t} = \dfrac{\pi}{\omega_2}$ into [4.43] leads to $\dot{u}_{2t}(\bar{t}) < 0$ because $\omega_1 < \omega_2$. This implies that $\tilde{t} < \bar{t} = \dfrac{\pi}{\omega_2}$ so $\dddot{u}_{2t}(\tilde{t}) < 0$ and $\dot{u}_{2t}(t) < 0\ \forall t \in]\tilde{t}, \tilde{t} + \eta[$. This contradicts the assumption that $\dot{u}_{2t}(t) > 0\ \forall t \in]\tilde{t}, \tilde{t} + \eta[$.

If conversely $\dot{u}_{2t}(t) < 0 \ \forall t \in \,]\tilde{t}, \tilde{t} + \eta[$, then mass 2 slides to the left and the motion of the two masses satisfies the following system:

$$\begin{cases} \ddot{u}_{1t} + C u_{1t} - u_{2t} = F_{1t} + \mu F_{1n} + \mu \varepsilon_1 \stackrel{def}{=} \tilde{F}_1, \\ \ddot{u}_{2t} - u_{1t} + C' u_{2t} = F_{2t} + \mu F_{2n} + \mu \varepsilon_2 \stackrel{def}{=} \tilde{F}_2. \end{cases} \qquad [4.44]$$

In this case, as opposed to the case when the tangential velocity of mass 2 was positive after $\tilde{t}$, the acceleration of mass 2 is a discontinuous function of time. When $CC' - 1 > 0$, the solution of this system is given by:

$$\begin{cases} u_{1t}(t) = \dfrac{C' \tilde{F}_1 + \tilde{F}_2}{CC' - 1} + c_1 (C' - \gamma_1^2) \cos \gamma_1 (t - \tilde{t}) + c_2 (C' - \gamma_1^2) \sin \gamma_1 (t - \tilde{t}) \\ \qquad\quad + c_3 (C' - \gamma_2^2) \cos \gamma_2 (t - \tilde{t}) + c_4 (C' - \gamma_2^2) \sin \gamma_2 (t - \tilde{t}) \\ u_{2t}(t) = \dfrac{C \tilde{F}_2 + \tilde{F}_1}{CC' - 1} + c_1 \cos \gamma_1 (t - \tilde{t}) + c_2 \sin \gamma_1 (t - \tilde{t}) \\ \qquad\quad + c_3 \cos \gamma_2 (t - \tilde{t}) + c_4 \sin \gamma_2 (t - \tilde{t}), \end{cases}$$

$$[4.45]$$

where

$$\gamma_1 = \sqrt{\dfrac{(C + C') - \sqrt{(C - C')^2 + 4}}{2}} \ \text{ and } \ \gamma_2 = \sqrt{\dfrac{(C + C') + \sqrt{(C - C')^2 + 4}}{2}}.$$

Then, equation [4.45] gives $\ddot{u}_{2t}^+(\tilde{t}) = -c_1 \gamma_1^2 - c_3 \gamma_2^2$. But whatever the sign of $CC' - 1$, the acceleration of mass 2 is given by:

$$\begin{aligned} \ddot{u}_{2t}^+(\tilde{t}) &= u_{1t}(\tilde{t}) - C' u_{2t}(\tilde{t}) - \frac{C' \tilde{F}_1 + \tilde{F}_2}{CC' - 1} + C' \frac{C \tilde{F}_2 + \tilde{F}_1}{CC' - 1} \\ &= u_{1t}(\tilde{t}) - C' u_{2t}(\tilde{t}) + \tilde{F}_2 \end{aligned} \qquad [4.46]$$

And finally

$$\ddot{u}_{2t}^+(\tilde{t}) = F_{2t} + \mu F_{2n} + \mu \varepsilon_2 + \bar{u}_{1t} - C' \bar{u}_{2t} + \frac{\mu(C \varepsilon_1 - \varepsilon_2 - C' \varepsilon_1 + CC' \varepsilon_2)}{C^2 - 1}$$
$$+ \frac{\mu}{2\omega_1^2}(1 - C')(\varepsilon_2 - \varepsilon_1) \cos \omega_1 \tilde{t} - \frac{\mu}{2\omega_2^2}(1 + C')(\varepsilon_2 + \varepsilon_1) \cos \omega_2 \tilde{t}, \qquad [4.47]$$

with

$$\bar{u}_{1t} = \frac{1}{cs}(F_{1n} + R_{1n}^{eq}), \quad \bar{u}_{2t} = -\frac{1}{cs}(F_{2n} + R_{2n}^{eq}).$$

From equation [4.47], the acceleration $\ddot{u}_{2t}^+(\tilde{t})$ can be written as:

$$\ddot{u}_{2t}^+(\tilde{t}) = \rho + \sigma\varepsilon_1 + \delta\varepsilon_2.$$

Considering $\ddot{u}_{2t}^+(\tilde{t})$ as a function of ε_1 and ε_2, it is immediately seen that

$$\rho > 0 \Longrightarrow \exists\varepsilon_0 > 0 \text{ such that } |\varepsilon_1| < \varepsilon_0 \text{ and } |\varepsilon_2| < \varepsilon_0 \Longrightarrow \ddot{u}_{2t}^+(\tilde{t}) > 0.$$

But

$$\rho = F_{2t} + \mu F_{2n} + \frac{1}{cs}(F_{1n} + R_{1n}^{eq}) + C'\frac{1}{cs}(F_{2n} + R_{2n}^{eq}), \qquad [4.48]$$

which can be written as:

$$\rho = \frac{1}{cs}(d_2 + R_{1n}^{eq} + CR_{2n}^{eq} - 2\mu csR_{2n}^{eq}) = -2\mu R_{2n}^{eq} > 0. \qquad [4.49]$$

Therefore, mass 2 stays motionless for some time after $\tilde{t}$. However, mass 1 continues to move and to affect the reaction of mass 2. Does the reaction of mass 2 stay strictly inside the Coulomb cone for all time following $\tilde{t}$? The computation of the tangential component of the reaction of mass 2 gives the answer. At $\tilde{t}$, the reaction of mass 2 is given by:

$$\tilde{R}_{2n} = -cs\tilde{u}_{2t} - F_{2n} - \varepsilon_2,$$

$$\tilde{R}_{2t} = -\tilde{u}_{1t} + (1 + c^2)\tilde{u}_{2t} - F_{2t}.$$

The reaction of mass 2 is strictly inside the Coulomb cone at time $\tilde{t}$, so that the following inequality holds:

$$\mu\tilde{R}_{2n} < \tilde{R}_{2t} < -\mu\tilde{R}_{2n}.$$

For $t > \tilde{t}$, the motion of mass 1 is given by:

$$u_{1t}(t) = \tilde{u}_{1t} + (\frac{U}{C} - \tilde{u}_{1t})(1 - \cos \sqrt{C}(t - \tilde{t})) + \frac{\tilde{\dot{u}}_{1t}}{\sqrt{C}} \sin \sqrt{C}(t - \tilde{t}),$$

where

$$U = F_{1t} + \mu F_{1n} + \mu \varepsilon_1 + \tilde{u}_{2t}, \quad \tilde{u}_{1t} = u_{1t}(\tilde{t}) \text{ and } \tilde{\dot{u}}_{1t} = \dot{u}_{1t}(\tilde{t}) < 0.$$

Therefore,

$$R_{2t} = \tilde{R}_{2t} - (\frac{U}{C} - \tilde{u}_{1t})(1 - \cos \sqrt{C}(t - \tilde{t})) - \frac{\tilde{\dot{u}}_{1t}}{\sqrt{C}} \sin \sqrt{C}(t - \tilde{t}),$$

but

$$\tilde{\dot{u}}_{1t} = \frac{\mu}{\omega_2}(\varepsilon_1 + \varepsilon_2) \sin \omega_2 \tilde{t},$$

and

$$\frac{U}{C} - \tilde{u}_{1t} = \frac{\mu}{2}(\varepsilon_1 - \varepsilon_2) \cos \omega_1 \tilde{t} + \frac{\mu}{2}(\varepsilon_1 + \varepsilon_2) \cos \omega_2 \tilde{t}.$$

So

$$|R_{2t} - \tilde{R}_{2t}| \le |\frac{\mu}{\omega_2 \sqrt{C}}(\varepsilon_1 + \varepsilon_2)| + |\mu(\varepsilon_1 - \varepsilon_2)| + |\mu(\varepsilon_1 + \varepsilon_2)|.$$

The conclusion is consequently that if ε_1 and ε_2 are sufficiently small the reaction of mass 2 always stays strictly inside the Coulomb cone, so that the motion of mass 1 can never put mass 2 back into motion. When in turn mass 1 stops the two masses are in a strictly stuck equilibrium. Once again this result supports conjecture 4.1.

4.5. Numerical experiments on a finite element discretization of an elastic body

This section contains a numerical computation of the trajectory of a system with a much larger number of degrees of freedom than the systems studied analytically in the preceding sections. The model is obtained by finite element discretization of a rectangular elastic block as presented in Chapter 2. There is of course no intention here to compute the dynamics of a continuous media with unilateral contact and Coulomb friction. Indeed in such a case, all the problems relating to continuum mechanics, in particular the convergence of the discretization, are unsolved and difficult problems. Therefore, the following computation must be viewed as a model generalizing, through the number of degrees of freedom, the one or two mass systems studied before. In fact, as mentioned in Chapter 2, the one mass models shown in Figure 1.4 have often been said to represent the behavior of a finite element mesh[1]. The computations are made using NSCD method (nonsmooth contact dynamics method) [JEA 99] implemented in the software $LMGC\,90$ (see [DUB 90]). A two-dimensional elastic block is meshed with 30×10 $Q4$ square elements, of length $l = 10^{-3}$ m. The material is linearly elastic in small perturbations with the following properties:

$$\left\{ \begin{array}{l} \text{mass per unit volume } \rho = 10^4 \text{ kg/m}^3, \\ \text{Young modulus } E = 5 \times 10^6 \text{ Pa}, \\ \text{Poisson ratio } \nu = 0.49, \\ \text{so that the celerity is } C = \sqrt{E/\rho} = 0.707\,10^2 \text{ m/s}. \end{array} \right. \qquad [4.50]$$

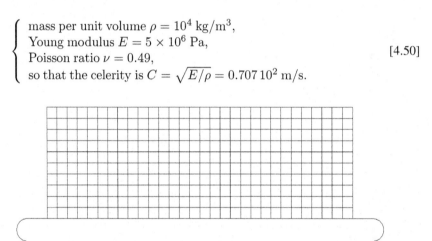

Figure 4.14. *The* 30×10 *$Q4$ meshed block*

The propagation time through a mesh element is $\tau = l/C = 0.14 \times 10^{-4}$ s. Planar deformations are assumed. The block is lying on the plane face of some fixed rigid object, referred to as the foundation (see Figure 4.14). The candidates to contact are numbered from 1 to 31 from left to right. Coulomb's friction is assumed between the block and the foundation with a friction coefficient $\mu = 1$. The collection of nodes of the upper layer is numbered from 1×11 to 31×11 from left to right. Each node

---

1 This section relies on a common work of the authors with Emeritus Researcher Michel Jean, mjean.recherche@wanadoo.fr, in particular all the numerical computations and postprocessing have been performed by Michel Jean.

numbered 8×11 to 24×11 is submitted to a vertical force of -0.75×10^3 N. The gravity forces are neglected [2].

4.5.1. A reference state with imminent sliding contacts

A dynamical computation is performed using a time step of 0.2×10^{-2} s, which is relatively larger than $\tau = 0.14 \times 10^{-4}$ s, so that a quasi-static evolution is computed practically instantly. The computation yields the equilibrium response shown in Figure 4.15 symmetric with respect to the vertical axis of the block (node 15, node 166).

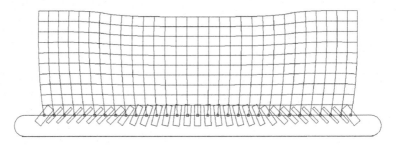

Figure 4.15. *The quasi-statically deformed block at equilibrium and the reaction forces at the nodes*

In this figure, reaction forces at contacting points are represented by rectangles, the longer side being directed with the reaction force, and the width of the rectangle being proportional to the force modulus $(R_t^2 + R_n^2)^{\frac{1}{2}}$. The nodes $2, 3, 4, 5, 6, 7, 8$, and $24, 25, 26, 27, 28, 29, 30$ are imminent sliding contacts (within an accuracy range of 10^{-3}).

The $|R_t|/\mu R_n$ distribution is displayed in Figure 4.16. This distribution is constructed as follows. Let $\chi \in [0, 1]$ be a real number and let N be the total number of contacts $^\alpha$ where μR_n^α is different from zero (here $N = 31$); let $N(\chi)$ be the number of contacts where $|R_t^\alpha|/\mu R_n^\alpha \geq \chi$; the $|R_t|/\mu R_n$ *distribution* is defined as the function $\chi \to N(\chi)/N$. This function is decreasing from 1 to some positive value, which is 0 if the sample is at equilibrium with reactions strictly inside the Coulomb cone. In this example, there are 14 imminent sliding contacts, i.e. a ratio of 0.45.

---

2 It might be observed that the Poisson ratio is very close to the limit of incompressibility. Indeed, this value, which may seem slightly artificial, has been chosen so as to force the lower boundary to remain in contact with the support while it is pushed in the direction tangent to the support.

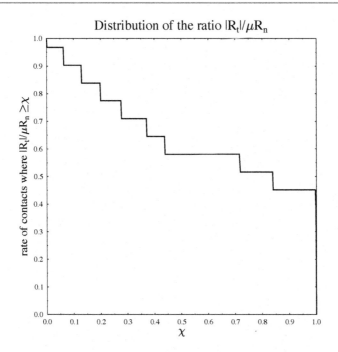

Figure 4.16. *The* $|R_t|/\mu R_n$ *distribution of the quasi-statically deformed block*

4.5.2. *The block perturbed by some horizontal shock*

Consider the block at the equilibrium state obtained by the quasi-static evolution above. A perturbation generated by the impact of a light rigid projectile is applied as shown in Figure 4.17. This rigid projectile is thrown on the left side of the block with a horizontal velocity equal to 3 m/s. The mass of the projectile is about 2.6% of the mass of the block.

Here, the time step is chosen so as to correctly capture dynamical evolutions. The time step is 0.1×10^{-5} s, which is therefore smaller than the time of propagation τ within a mesh element $\tau = l/C = 0.14 \times 10^{-4}$ s. After the episode of impact, the projectile is thrown backward and a complex system of waves is generated (see Figure 4.20), displaying the nodes velocity field at 0.2×10^{-3} s after the projectile has bounced off the block. It may be observed that the amplitude of the waves is decaying near the bottom of the block in contact with the foundation. A pure longitudinal wave would take approximately 0.84×10^{-3} s to propagate back and forth through the block and the numerical simulation lasts 0.8×10^{-1}. By this time, waves are vanishing slowly because of some light numerical damping implemented in the algorithm equivalent to a Rayleigh internal damping. One may estimate that the impact is mild, in the sense that the middle contacting node 16 does not move during the experiment. The final distribution of reaction forces is shown in Figures 4.18 and 4.19. Figure 4.18 is not significantly different from Figure 4.15. However, the distribution of forces is no longer symmetric as it was in the quasistatically deformed block. In Figure 4.19, the dotted line represents the $|R_t|/\mu R_n$ distribution for the quasi-static evolution given in Figure 4.16, and the thick line represents the distribution after impact. The figure shows that imminent slidings have been destroyed and all the reactions on the contact boundary are now strictly inside the cone, though some contacts are still close to sliding. When tracking the status of contacting nodes during the experiment, it may be seen that some imminent sliding contacts slide back and forth before sticking as observed in many cases on simpler models. In this experiment, the amplitude of the waves is relatively smaller near the contact zone where the body is stuck than in the upper layers where the body is free. Nevertheless, microtraveling waves near the contact zone change the status of the contacting nodes to finally destroy all imminent slidings. We may imagine that milder perturbations may destroy only a subcollection of imminent slidings.

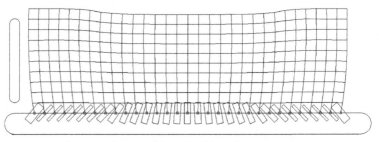

Figure 4.17. *The quasi-statically deformed block at equilibrium, and some rigid projectile ready to be thrown horizontally to hit the left side of the block*

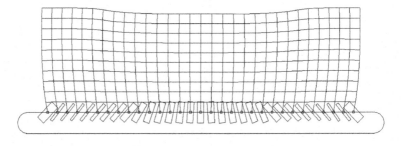

Figure 4.18. *The settled deformed block after impact. The projectile has been sent back out of the frame*

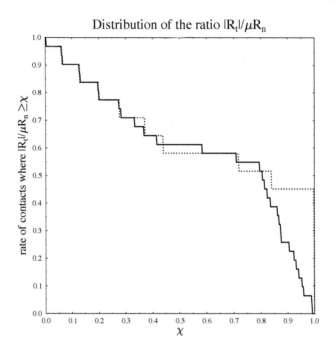

Distribution of the ratio $|R_t|/\mu R_n$

rate of contacts where $|R_t|/\mu R_n \geq \chi$

χ

Figure 4.19. *The* $|R_t|/\mu R_n$ *distribution of the settled impacted block*

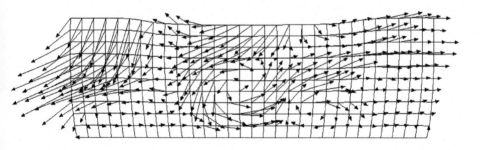

Figure 4.20. *The nodes velocity field at* 0.2×10^{-3} s
after the projectile has bounced off the block

This experiment emphasizes the fact, which we reluctantly admit, that the set of reactions ensuring the equilibrium of a structure under a given loading is far from unique. Indeed, two different distributions of the reaction forces equilibrating the loading have been obtained in the present case: the first one is issued from a quasi-static evolution and the second is obtained after a dynamical evolution. This numerical experiment furnishes a further justification of the conjecture. Indeed, all the behaviors described in the previous sections have been observed here.

The main feature is clearly observed: after perturbing an equilibrium state, the trajectory leads to a new state where all the reactions are strictly inside the cone. This means, from a mechanical point of view, that starting from an imminent sliding state, the final state is strictly stuck by friction.

The points which are not in contact but which are only connected to points in contact by an elastic stiffness continue to oscillate indefinitely, so a final state where all the points are at equilibrium can be reached only by introducing some damping.

5

Exploring the Case of the Linear Restoring Force

This chapter investigates the dynamics of the simple mass-spring system with a linear restoring force involving non-regularized unilateral contact and Coulomb friction. In Chapter 4, the set of equilibria was shown to have very specific features in comparison to those of classical dynamical systems. Indeed, as will be discussed throughout this chapter, the dynamics of this system involving Coulomb friction is far from classical.

The question relating to the qualitative analysis of dynamical systems is as follows: when a system involving unilateral contact and dry friction is submitted to an oscillating excitation, can its answer be predicted by partitioning the {$period$, $amplitude$} plane of the excitation into different kinds of qualitative behaviors? The answer is yes. The {$period$, $amplitude$} plane is in fact shown to be essentially divided into three ranges. One range, for external forces of sufficiently small amplitude, is a horizontal strip, where no periodic solutions exist but where any point corresponds to infinitely many equilibrium states. The second range is where no equilibrium states exist but the whole range contains periodic solutions, the number and the period of which depend on both the period of the excitation and its amplitude. The second range has a relatively intricate upper boundary above which the third range lies, where periodic solutions still exist, but where all initial data lead to a trajectory which loses contact, so the periodic trajectories involve jumps and shocks. It is important to stress the fact that this analysis was possible due to a rigorous statement of the dynamics (see problem [2.1] and section 3.1) and to the proof that the corresponding Cauchy problem was well-posed.

The transition between the range where there exist only equilibrium states and no periodic solutions and the range where there exist periodic solutions and no equilibrium states is of particular interest because of the variety of different behaviors of the solution. A thorough analysis of this transition is very important from the point of view of engineering applications. Indeed, passing through the transition in the present case means that under loads of increasing amplitude, all the equilibrium states disappear while the system undergoes a sudden onset of oscillating regimes. In engineering terms, a kind of intrinsic stabilization or control evolves into permanent vibrations. Moreover, the investigation of this occurrence will show that it is much more intricate than classical Hopf bifurcations, one of the main reasons being the strong dependence on the frequency of the excitation. Therefore, an accurate understanding of how the transition occurs should be useful to ensure the safe use of any machine involving rotating or sliding mechanisms.

The dynamical problem studied throughout this chapter was established in Chapter 3 and is written as follows:

$$
\begin{cases}
i) \begin{cases} m\ddot{u}_t + K_t u_t + W u_n = F_t + R_t, \\[6pt] m\ddot{u}_n + W u_t + K_n u_n = F_n + R_n, \end{cases} \quad t > 0 \\[14pt]
ii) \ u_t(0) = u_{t0}, \ u_n(0) = u_{n0}, \ \dot{u}_t(0) = v_{t0}, \ \dot{u}_n(0) = v_{n0}, \\[10pt]
iii) \ u_n \leq 0, \ R_n \leq 0, \ u_n R_n = 0, \\[10pt]
iv) \begin{cases} R_n = 0 \implies \dot{u}_t \in \mathbb{R}, \\[6pt] \mu R_n \leq R_t \leq -\mu R_n, \\[6pt] \text{with} \begin{cases} |R_t| < -\mu R_n \implies \dot{u}_t = 0, \\[4pt] |R_t| = -\mu R_n \implies \exists \lambda \geq 0 \text{ s.t. } \dot{u}_t = -\lambda R_t, \end{cases} \end{cases} \\[22pt]
v) \ u_n(t) = 0 \implies \exists e \in [0,1] \text{ such that } \dot{u}_n^+(t) = -e\dot{u}_n^-(t).
\end{cases}
\tag{5.1}
$$

The investigation of the equilibrium states of this problem, carried out in Chapter 4, can be summarized as follows:

– the structure of the set of equilibria depends only on the sign of two quantities that are $A = K_t F_n - W F_t$ and $\mu - \frac{K_t}{W}$;

– equilibrium states without contact always exist when $A < 0$. The equilibrium state without contact is then unique. Moreover, if $\mu - \frac{K_t}{W} > 0$, then the equilibrium state without contact coexists with infinitely many equilibrium states in contact with a strictly negative normal component of the reaction;

– if $A \geq 0$, there are no equilibrium states without contact. The set of equilibria is reduced to a single state in grazing contact if $A = 0$ and $\mu - \frac{K_t}{W} < 0$;

– if $A > 0$, there always exist infinitely many equilibrium states, which all have a strictly negative normal component of the reaction. If in addition $\mu - \frac{K_t}{W} \geq 0$, then this set fills completely a half-line in the $\{R_t, R_n\}$ plane, while it fills only a bounded interval if $\mu - \frac{K_t}{W} < 0$. In this case, the ends of the half-line or of the interval are in imminent sliding, while all the other equilibria are strictly stuck by friction.

By far the most interesting situation observed under a constant external force is the one corresponding to the choice of parameters ensuring that the equilibria are all in strict contact (i.e. no grazing contact) and fill an interval in the $\{R_t, R_n\}$ plane (i.e. $\mu W < K_t$ and $A = K_t F_n - W F_t > 0$), as represented in Figure 5.1. This case is the most general in the sense that it involves all the difficulties that can be encountered separately in the other cases. All the equilibrium states are strictly stuck by friction except the two ends of the interval that are in imminent sliding, one to the right and the other to the left. The analysis presented here consists of submitting the system to an additional oscillating force, and in studying the effects of this perturbation on the equilibria.

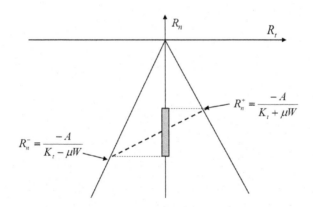

Figure 5.1. *The set of equilibrium states, for $A > 0$ and $\mu - \frac{K_t}{W} < 0$. It is given by the intersection of the Coulomb cone with a manifold, here a straight line, which represents the equilibrium equation*

5.1. Equilibrium states in the $\{R_t, R_n\}$ plane under an oscillating loading

As stated in Chapter 5, if one of the strictly stuck equilibria represented in Figure 5.1 is perturbed by a sufficiently small additional force, the mass remains at rest as long as the perturbation does not bring the reaction to the border of the cone. As such, it is immediately seen that for any given loading $(F_t(t), F_n(t))$, the set of the normal component of the reactions corresponding to equilibrium solutions is an interval depending on time and given by:

$$\{R_n\}(t) = \left[\frac{-K_t F_n(t) + W F_t(t)}{K_t - \mu W}, \frac{-K_t F_n(t) + W F_t(t)}{K_t + \mu W} \right].$$ [5.2]

For any given t, the set $\{R_n\}(t)$ is the interval of the R_n axis represented by a thick line in Figure 5.1. The following result will be a fundamental tool for a large part of the analysis.

LEMMA 5.1.– Let A be positive, the loading be piecewise analytical and let $\{R_n\}(t)$ be the set of normal components of the reactions of the equilibria at time t. Then, if the trajectory of a sliding mass that satisfies problem [5.1] is such that at instant t^* when the mass stops sliding, the normal component of the reaction R_n^* belongs to the interior of $\{R_n\}(t^*)$, then the mass will remain in a strictly stuck equilibrium state as long as the normal component of the reaction remains in the interior of $\{R_n\}(t)$.

PROOF.– Existence and uniqueness results for problem [5.1] have been given in Chapter 3 and in particular uniqueness holds as soon as the loading is piecewise analytical. As such, any continuous function that satisfies all the relations in [5.1] will be the unique solution of [5.1].

Let $(u_t(t), R_t(t), R_n(t))$ be a solution of [5.1] in $[0, t^*]$ strictly in contact (i.e. $u_n(t) \equiv 0$) such that $\dot{u}_t(t^*) = 0$ and R_n^* belongs to the interior of the set $\{R_n\}(t^*)$. Then, this solution can be extended for $t^* < t < \bar{t}$ by:

$$u_t(t) = u_t(t^*), \dot{u}_t(t) = 0, R_t(t) = K_t u_t(t^*) - F_t(t)$$

and

$$R_n(t) = W u_t(t^*) - F_n(t).$$

Time $\bar{t}$ stands for the first instant for which $R_n(t)$ does not belong to the set $\{R_n\}(t)$. If $R_n(t)$ belongs to the interior of $\{R_n\}(t)$ for all t, then the solution is a strictly stuck equilibrium solution.   □

REMARK 5.1.– This lemma generalizes lemma 4.1 in which the loading was constant.

In order to study the response of the system to an oscillating loading, the functions $F_t(t)$ and $F_n(t)$ are now written in the following way:

$$F_t(t) = F_t + P_t(t) \quad \text{and} \quad F_n(t) = F_n + P_n(t),$$

where F_t and F_n are constant and $P_t(t)$ and $P_n(t)$ are tangential and normal perturbations.

This study focuses on the case of a tangential perturbation, in other words on the case $P_n(t) = 0$. It can indeed be easily verified that, due to the coupling, the qualitative analysis is the same in the case of a perturbation with a non-zero normal component. In fact, adding a perturbation to the normal component increases the set $\{R_n\}(t)$ of normal components of the reaction corresponding to stationary solutions so that the analysis loses much of its interest.

In order to perform closed-form calculations as far as possible, the oscillating perturbation considered is a tangential perturbation of rectangular wave shape defined in the following way:

$$\begin{cases} \text{For } i = 0, 1, 2, \dots : \\ P_t(t) = \varepsilon \text{ if } t \in \,]2iT, (2i+1)T], \\ P_t(t) = 0 \text{ if } t \in \,](2i+1)T, (2i+2)T]. \end{cases} \qquad [5.3]$$

Therefore, $\varepsilon > 0$ is the amplitude and T the half-period of the perturbation. In Figure 5.2, the equilibrium solutions corresponding to $P_t(t) = 0$ and to $P_t(t) = \varepsilon$ are represented in the $\{R_t, R_n\}$ plane. The set $\overline{R}_n$ represents the set of equilibrium solutions under the loading $(F_t(t), F_n(t))$:

$$\overline{R}_n = \bigcap_{t>0} \{R_n\}(t).$$

In the case of the tangential rectangular wave shape chosen here $(F_t(t) = F_t + P_t(t), F_n(t) = F_n)$, $\overline{R}_n$ becomes:

$$\begin{aligned} \overline{R}_n &= \left[\frac{-K_t F_n + W F_t}{K_t - \mu W}, \frac{-K_t F_n + W F_t}{K_t + \mu W} \right] \\ &\quad \cap \left[\frac{-K_t F_n + W(F_t + \varepsilon)}{K_t - \mu W}, \frac{-K_t F_n + W(F_t + \varepsilon)}{K_t + \mu W} \right], \end{aligned}$$

$$\overline{R}_n = \left[\frac{-A + \varepsilon W}{K_t - \mu W}, \frac{-A}{K_t + \mu W} \right]. \qquad [5.4]$$

The following notations are now introduced as these particular values of the normal component of the reaction will appear very often from now on:

$- R_n^- = \dfrac{-A}{K_t - \mu W}$ and $R_n^+ = \dfrac{-A}{K_t + \mu W}$, which correspond, respectively, to imminent sliding to the right and to the left during the half-periods where $P_t(t) = 0$;

$- R_{n\varepsilon}^- = \dfrac{-A + \varepsilon W}{K_t - \mu W}$ and $R_{n\varepsilon}^+ = \dfrac{-A + \varepsilon W}{K_t + \mu W}$, which correspond, respectively, to imminent sliding to the right and to the left during the half-periods where $P_t(t) = \varepsilon$.

The expression of $\overline{R}_n$ given by equation [5.4] can then be written:

$$\overline{R}_n = \left[R_{n\varepsilon}^-, R_n^+\right].$$

This set $\overline{R}_n$ can be an interval on the R_n axis (represented by a thick line in Figure 5.2), and can be reduced to a single point or can be empty depending on the value of ε. The following theorem can be deduced from equation [5.4].

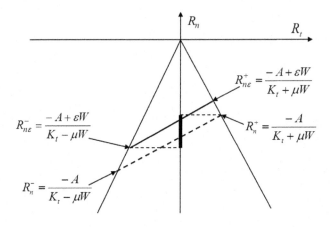

Figure 5.2. *Equilibrium solutions in the $\{R_t, R_n\}$ plane (dotted line: equilibrium solutions when the tangential perturbation is equal to zero, full line: equilibrium solutions when the tangential perturbation is equal to ε)*

THEOREM 5.1.– The existence of equilibrium solutions depends on the value of ε in the following way:

– when $\varepsilon > \dfrac{2\mu A}{K_t + \mu W}$, the set of normal components of the reactions at equilibrium $\overline{R}_n$ is an empty set so there are no equilibrium solutions;

– when $\varepsilon = \dfrac{2\mu A}{K_t + \mu W}$, the set $\overline{R}_n$ reduces to a single point so there is a unique equilibrium solution;

– when $\varepsilon < \dfrac{2\mu A}{K_t + \mu W}$, the set $\overline{R}_n$ is a non-zero measure interval so there exist infinitely many equilibrium solutions.

To illustrate these results, numerical calculations will be performed with the Maple software, using the following values for the data:

$$m = 1, \ \mu = 0.5 \text{ and } Fn = 2, \ Ft = 1, \ Kt = 2, \ W = 1 \text{ so that } A = 3. \quad [5.5]$$

For example, $\frac{2\mu A}{K_t + \mu W} = 1.2$ when the above values of the data are used. As the parameter m has no influence on this study, m will be taken equal to 1 in all the following results whether numerical or theoretical.

5.2. When no equilibrium solutions exist

The range where $\varepsilon > \frac{2\mu A}{K_t + \mu W}$ in the $\{T, \varepsilon\}$ plane is divided into two parts separated by a relatively complicated boundary: a "lower" range where the particle is always in contact, and an "upper" range, where the trajectory loses contact at least once. In the range where the trajectories are always in contact, the existence of periodic solutions will be proved and a detailed description of the different shapes taken by the orbits will be given. Then, special attention will be paid to the upper boundary of this range defined as the value of ε corresponding to the occurrence of the loss of contact. This boundary will be shown to depend strongly on the period of the excitation. Trajectories that lose contact, therefore involving impacts, will be studied at the end of this section.

Periodic solutions of different kinds and multiplicity exist for any value of the half-period of the excitation T as soon as there no longer exist equilibrium states. Due to the nonsmoothness of system [5.1], the existence of periodic solutions cannot result from classical theories of dynamical systems. The existence of these periodic solutions will be established in some cases by reasoning in the reaction plane and in others by following the solution by piecewise calculations for given values of the parameters and proving that there exist solutions for which the trajectory comes back to its initial data. Of course, this ensures the existence of some periodic solutions, but does not prove that there do not exist others. Nevertheless, in some cases, the well-posedness of the Cauchy problem will exclude the existence of other periodic solutions. The mechanical parameters will be fixed throughout the analysis, with the explicit values given in [5.5] for all the numerical computations, and the $\{period, amplitude\}$ plane will be explored. The numerical experiments given here not only illustrate the theoretical results but are also often used as a guide for the calculations that were subsequently performed to prove these results (see [GEY 00] for example).

5.2.1. Sliding periodic solutions

Although the dynamical problem [5.1] describes the dynamics in any situation, it will be useful to specify the equations of the trajectories during sliding phases (see

Chapter 3, section 3.2.3). Inserting $u_n \equiv 0$ into equation [5.1] together with $R_t = \mu R_n$ yields the equation of the motion when sliding to the right:

$$\ddot{u}_t + (K_t - \mu W)u_t = F_t - \mu F_n, \qquad\qquad [5.6]$$

while $R_t = -\mu R_n$ yields the equation of the motion when sliding to the left:

$$\ddot{u}_t + (K_t + \mu W)u_t = F_t + \mu F_n. \qquad\qquad [5.7]$$

It is interesting to observe that, although they strictly obey Coulomb's friction law, the reactions have been eliminated from each equation so that each sliding phase is a part of a linear oscillation.

The half-periods of the free oscillations respectively denoted by T_α and T_β for sliding to the right and to the left, are consequently given by:

$$T_\alpha = \frac{\pi}{\sqrt{K_t - \mu W}}, \quad T_\beta = \frac{\pi}{\sqrt{K_t + \mu W}}$$

The corresponding frequencies ω_α and ω_β already introduced in Chapter 5 are given by $\omega_\alpha^2 = K_t - \mu W$ and $\omega_\beta^2 = K_t + \mu W$.

The analysis will be divided into several ranges of the half-period T, $T > T_\alpha + T_\beta$ where, in particular, infinitely many periodic solutions of period $4T$ will be found, $T_\beta < T < T_\alpha + T_\beta$ that will appear as relatively intricate, and $T < T_\beta$.

5.2.1.1. Periodic solutions for large T

The proof of the existence of periodic solutions in this range where $T > T_\alpha + T_\beta$ is based upon the following remarks:

– since the period is sufficiently large, the explicit solution in each part of the trajectory is the same as if the loading were a constant;

– in each given sliding phase with initial data $\{u_0, v_0\}$ with u_0 out of equilibrium and $v_0 = 0$, the sliding motion has a first stopping at a point u_1, symmetrical of u_0, with respect to the equilibrium point in imminent sliding in the same direction, which is nothing but the elementary property of a linear oscillator with constant right-hand side and constant coefficients;

– moreover, u_t or R_n can indifferently be used as the basic unknown of the analysis. Indeed for a sliding motion, the position u_t is connected to the normal component of the reaction R_n by the second equation of system [5.1(i)], which, because u_n is identically equal to zero, reduces to: $W u_t = F_n + R_n$. To this must be added the fact that, as the period of the excitation is sufficiently large, the particle always attains a zero velocity after a sliding phase before being set into motion again.

A periodic solution in the $\{R_t, R_n\}$ plane can be built in the following way (see Figure 5.3). Let $R_n^{\varepsilon-}$ be the normal component of the reaction of the equilibrium state in imminent sliding to the right when $P_t(t) = \varepsilon$, and $R_n^{\varepsilon-} - x$ (with $x > 0$) is taken as initial data with zero velocity. T larger than T_α implies that the velocity goes through zero at the point $R_n^{\varepsilon-} + x$. Assume in addition that x is such that $R_n^{\varepsilon-} + x$ is larger than $R_n^{\varepsilon+}$, where $R_n^{\varepsilon+}$ is the reaction of the equilibrium state in imminent sliding to the left still with $P_t(t) = \varepsilon$, and let $R_n^{\varepsilon-} + x = R_n^{\varepsilon+} + y$, $y > 0$. Then, the reaction jumps to the other side of the cone and the particle slides to the left during a half-period T_β up to a stop since $T > T_\alpha + T_\beta$. The reaction then jumps to an equilibrium state, with $R_n = R_n^{\varepsilon+} - y$ and stays there until the change of the external load at time T, when $P_t(t)$ is set to zero. Then, let $R_n^{\varepsilon+} - y = R_n^+ + z$, $z > 0$. At time T, the particle is set into motion again and slides to the left during a period T_β up to a stop at the reaction $R_n^+ - z$ and then jumps to an equilibrium and stays there until the time $2T$ where the data will be set to $P_t(t) = \varepsilon$ again. Evidently, such a trajectory is periodic with period $2T$ if the condition $R_n^+ - z = R_n^{\varepsilon-} - x$ is satisfied. Substituting z and y in this periodicity condition gives an equation for the initial data:

$$
\begin{aligned}
R_n^{\varepsilon-} - x &= R_n^+ - z \\
&= 2R_n^+ - R_n^{\varepsilon+} + y \\
&= 2R_n^+ - 2R_n^{\varepsilon+} + R_n^{\varepsilon-} + x,
\end{aligned}
\qquad [5.8]
$$

so

$$
x = R_n^{\varepsilon+} - R_n^+ = \frac{\varepsilon W}{K_t + \mu W}. \qquad [5.9]
$$

This trajectory exists if it is compatible with the description given above, in other words if $0 \le y \le R_n^{\varepsilon+} - R_n^{\varepsilon-}$. Finally, this unique periodic solution exists only when $\frac{2\mu A}{K_t + \mu W} \le \varepsilon \le \frac{4\mu A}{K_t + 3\mu W}$. This trajectory is represented in Figure 5.4(a) in the $\{R_t, R_n\}$ plane, and in Figure 5.4(b) in the phase space, where it can be observed that because of the successive stops, the trajectory is not diffeomorphic to an ellipse. For a given value of the pair (u_0, v_0), due to the well-posedness of the Cauchy problem, no other trajectory passing through this initial data exists.

In a similar way, if the periodicity condition $R_n^+ - z = R_n^{\varepsilon-} - x$ is removed, a periodic solution of period $4T$ can be obtained by doing two complete loops instead of one. Using the same definition as above for x, y and z and introducing the quantities: $w = R_n^{\varepsilon-} - (R_n^+ - z)$, $q = R_n^{\varepsilon-} + w - R_n^{\varepsilon+}$ and $u = R_n^{\varepsilon+} - q - R_n^+$ for the second loop, a periodic solution of period $4T$ exists if and only if $R_n^+ - u = R_n^{\varepsilon-} - x$. A simple computation shows that the periodicity condition is satisfied for any compatible value of x, that is for any value of x implying that the values of y, z and q are compatible with the above construction. As such, this time, not only must $0 \leq y \leq R_n^{\varepsilon+} - R_n^{\varepsilon-}$ but $0 \leq z \leq R_n^+ - R_n^-$ and $y \leq q \leq R_n^{\varepsilon+} - R_n^{\varepsilon-}$ must also be satisfied. By a straightforward but rather tedious computation, these three conditions are shown to imply that all values of x belonging to a certain interval correspond to a periodic solution of period $4T$. If $K_t \leq 5\mu W$, this interval is given by:

$$\max(\mathrm{R}_n^{\varepsilon+} + \mathrm{R}_n^{\varepsilon-}, 2(\mathrm{R}_n^{\varepsilon-} + \mathrm{R}_n^+)) \leq \mathrm{x} \leq \mathrm{R}_n^{\varepsilon+} + \mathrm{R}_n^+. \qquad [5.10]$$

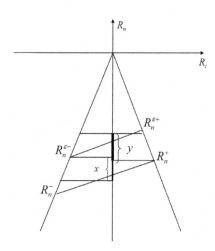

Figure 5.3. *Definition of the quantities x and y*

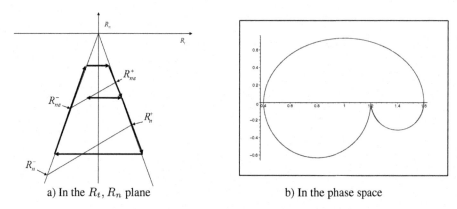

a) In the R_t, R_n plane   b) In the phase space

Figure 5.4. *Trajectories of solutions of period $2T$ for $\varepsilon = 1.5$ and $T = 4.6$ (larger than $T_\alpha + T_\beta$)*

The case $K_t \geq 5\mu W$ would lead to a smaller interval obtained by a similar calculation.

In both cases, this interval of possible values for x is not empty as long as:

$$\frac{2\mu A}{K_t + \mu W} < \varepsilon < \frac{4\mu A}{K_t + 3\mu W}.$$

Thus, the following result has been established:

PROPOSITION 5.1.– For all $T \geq T_\alpha + T_\beta$,

– if $\varepsilon \in \left[\dfrac{2\mu A}{K_t + \mu W}, \dfrac{4\mu A}{K_t + 3\mu W} \right]$, there exists a unique periodic solution of period $2T$ obtained with the initial data $u_0 = (F_n + R_n^{\varepsilon-} - R_n^{\varepsilon+} + R_n^+)/W$ and $v_0 = 0$;

– if $\varepsilon \in \left] \dfrac{2\mu A}{K_t + \mu W}, \dfrac{4\mu A}{K_t + 3\mu W} \right[$, there exists infinitely many periodic solutions of period $4T$ obtained with any initial data u_0 in a bounded non-zero measure interval and $v_0 = 0$.

REMARK 5.2.–

i) $K_t > \mu W$ implies that the interval $\left] \dfrac{2\mu A}{K_t + \mu W}, \dfrac{4\mu A}{K_t + 3\mu W} \right[$ is not empty;

ii) If, for example, $K_t \leq 5\mu W$, the non-zero measure interval of proposition 5.1 is:

$$](F_n + R_n^{\varepsilon-} - R_n^{\varepsilon+} + R_n^+)/W, (F_n + R_n^{\varepsilon-} - \max(R_n^{\varepsilon+} + R_n^{\varepsilon-}, 2(R_n^{\varepsilon-} + R_n^+))/W];$$

iii) The lower bound of this interval corresponds to the unique periodic solution of period $2T$;

iv) If for any pair (ε, T) in the range studied in proposition 5.1, initial data that do not belong to the non-zero measure interval are chosen, numerical observation implies that the solution is not periodic but converges to one of the periodic solutions of period $4T$ in finite time.

Examples of trajectories of period $4T$ are represented in Figure 5.5.

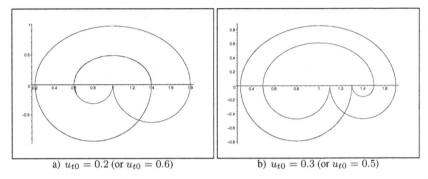

a) $u_{t0} = 0.2$ (or $u_{t0} = 0.6$)   b) $u_{t0} = 0.3$ (or $u_{t0} = 0.5$)

Figure 5.5. *Trajectories in the phase space* $(u_t, \dot{u}_t)$ *of solutions of period* $4T$ *for* $\varepsilon = 1.5$ *and* $T = 4.6$ *(larger than* $T_\alpha + T_\beta$*)*

5.2.1.2. *Investigation for decreasing values of* T

When T becomes smaller than $T_\alpha + T_\beta$, the qualitative behavior of the oscillator becomes more intricate than for larger values of T. In fact, the $\{T, \varepsilon\}$ plane must be divided into different zones and in each of these zones, the oscillator behaves differently. Figure 5.6 represents the central part of the $\{T, \varepsilon\}$ plane (see Figure 5.7) where six zones appear that are described qualitatively in the following:

Ω_1 : The periodic solutions represented in Figures 5.4 and 5.5 no longer exist when $T < T_\alpha + T_\beta$, but the existence of periodic solutions of period $2T$ when T is close to $T_\alpha + T_\beta$ will nevertheless be proved by a direct analytical calculation.

Ω_2 : A qualitative change appears on the left of Ω_1 by the occurrence of a phase difference between the loading and the periodic solutions.

Ω_3 : The left boundary of Ω_2 corresponds to the loss of existence of sliding solutions, so that Ω_3 contains solutions that lose contact. This loss of contact can occur even for small values of ε.

Ω_4 : This region is bounded on the left by the axis $T = 0$ and contains periodic solutions that are also out of phase with respect to the loading.

Ω_5 and Ω_6 : Between Ω_2 and Ω_4 are found very small zones referred to as Ω_5 and Ω_6 that contain other types of periodic solutions. Locally, the structure of the $\{T, \varepsilon\}$ plane is relatively intricate but going through Ω_5 and Ω_6, the transition from Ω_2 to Ω_4 appears as a continuous deformation of the orbits.

With logical notations, Γ_{ij} will denote the common boundary between Ω_i and Ω_j.

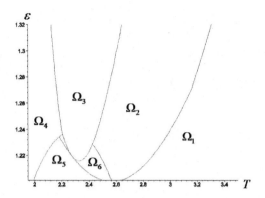

Figure 5.6. *A zoom, with an important vertical expansion, of the central part inside the rectangle that appears in Figure 5.7*

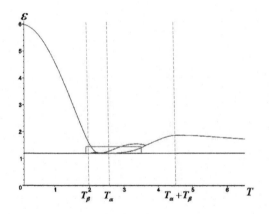

Figure 5.7. *The global period/amplitude plane*

5.2.1.2.1. Zone Ω_1 : T close to $T_\alpha + T_\beta$

In this range of half-periods of the excitation smaller than those studied in the previous section, the case of T sufficiently close to $T_\alpha + T_\beta$ will be of special interest since, although the previous simple analysis no longer applies, the existence of a single periodic solution is still established with an explicit formula for the corresponding initial data. The limit of the range in the $\{T, \varepsilon\}$ plane, where these solutions exist, is then also obtained explicitly. Moreover, new qualitative changes appear in the trajectory as T decreases. The tangential position u_t is now used as the unknown of the dynamical problem, instead of R_n, these two unknowns being linked by $W u_t = F_n + R_n$ as stated above. Let $u^{\varepsilon-}$ and u^- be the positions of the equilibrium state in imminent sliding to the right, respectively for $P_t(t) = \varepsilon$ and for $P_t(t) = 0$, and let the initial data be $\{u_0, v_0\} := \{u^{\varepsilon-} - x, 0\}$ for some unknown positive x. Since $T > T_\alpha$, the first part of the trajectory leads to a first stop at $u_1 = u^{\varepsilon-} + x$ for $t = T_\alpha$ where there is a jump of the reaction to the other side of the cone. From this time onward, the remaining part of the trajectory is the solution to the following system:

$$\begin{cases} \ddot{u}_2 + \omega_\beta^2 u_2 = F_t + \mu F_n + \varepsilon, \ t \in (T_\alpha, T) \\ u_2(T_\alpha) = u_1 = u_{\varepsilon-} + X, \ \dot{u}_2(T_\alpha) = 0, \\[2mm] \ddot{u}_3 + \omega_\beta^2 u_3 = F_t + \mu F_n, \ t \in (T, \tilde{t}) \\ u_3(T) = u_2(T), \ \dot{u}_3(T) = \dot{u}_2(T), \\[2mm] \tilde{u} = u_3(\tilde{t}), \ \tilde{t} \text{ such that } \dot{u}_3(\tilde{t}) = 0. \end{cases} \qquad [5.11]$$

Adding the periodicity condition given by $\tilde{u} = u_0$, one and only one value of x given by the following explicit formula is obtained:

$$x = \frac{\dfrac{\varepsilon}{\omega_\beta^2} \left[u^{\varepsilon-} - u^{\varepsilon+}\right] (1 - \cos\omega_\beta(T - T_\alpha))}{2\left[u^{\varepsilon-} - u^{\varepsilon+}\right] + \dfrac{\varepsilon}{\omega_\beta^2}(1 + \cos\omega_\beta(T - T_\alpha))}, \qquad [5.12]$$

where $u^{\varepsilon-} = \dfrac{F_t - \mu F_n + \varepsilon}{\omega_\alpha^2}$ and $u^{\varepsilon+} = \dfrac{F_t + \mu F_n + \varepsilon}{\omega_\beta^2}$.

Due to the well-posedness of the Cauchy problem, no other trajectory passes through the point $(u_0, 0)$.

From formula [5.12], the periodic solution of period $2T$ is obtained for any ε in the corresponding range and T smaller than, but close to, $T_\alpha + T_\beta$. Using the numerical values chosen in equation [5.5], an example of such a periodic solution is given in Figure 5.8(a) in the case $T = T_\alpha + \frac{2}{3}T_\beta$.

What happens for decreasing values of T? The use of problem [5.11] for the calculation of a periodic solution relies on the fact that the particle remains at rest at the point $\{\tilde{u}, 0\}$ up to the time when the perturbation of the loading is set to $P_t(t) = \varepsilon$ again. If $\tilde{u} < u^-$, the point $\{\tilde{u}, 0\}$ is not an equilibrium point for $P_t(t) = 0$, so the reaction of the particle passing through this point can only jump to the other side of the cone and the particle starts sliding in the other direction. In other words, formula [5.12] no longer applies because problem [5.11] itself no longer applies. Formula [5.12] establishes that x is a continuously decreasing function of T

in the interval $]T_\alpha, T_\alpha + T_\beta[$, so that there exists in this interval a single value of T depending on ε for which the initial data u_0 are such that $u_0 = u^-$. Let T_0 be this value connected to ε by the following equation:

$$T_0 - T_\alpha - \frac{1}{\omega_\beta} \arccos \frac{u^{\varepsilon-} - u^{\varepsilon+} - \left[\frac{2\omega_\beta^2}{\varepsilon}\left(u^{\varepsilon-} - u^{\varepsilon+}\right) + 1\right]\left(u^{\varepsilon-} - u^-\right)}{2u^{\varepsilon-} - u^- - u^{\varepsilon+}} = 0. \quad [5.13]$$

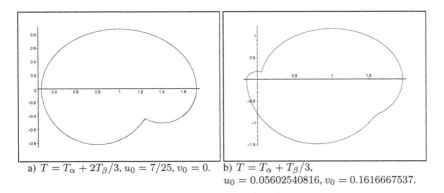

a) $T = T_\alpha + 2T_\beta/3$, $u_0 = 7/25$, $v_0 = 0$.   b) $T = T_\alpha + T_\beta/3$,
$u_0 = 0.05602540816$, $v_0 = 0.1616667537$.

Figure 5.8. *The difference between periodic solutions on both sides of* Γ_{12} *for* $\varepsilon = 1.5$

An explicit expression of the boundary Γ_{12} is thus obtained. Problem [5.11] could be modified in order to take one more sliding phase into account, but this would not give a periodic solution of period $2T$ starting with a zero velocity at time $t = 0$ if $\tilde{u} < u^-$, that is if T is on the left of boundary Γ_{12}. The reason is that there does not remain enough time for a complete new sliding phase to occur up to an equilibrium point before $t = 2T$.

Nevertheless periodic solutions still exist, but they do not start with a zero velocity at the origin. This means that the left side of zone Ω_1 will be characterized by the occurrence of a phase difference between the periodic loading and the response (see Figure 5.8(b)).

5.2.1.2.2. Zone Ω_2 : periodic solutions on the left of Ω_1

As a result of the analysis of zone Ω_1, when (T, ε) belongs to Ω_2, periodic solutions are sought for starting at time $t = 0$ from a point $\{u_0, v_0 \neq 0\}$. The trajectory can be qualitatively described as follows:

– the particle is sliding to the right (e.g. with a positive initial tangential velocity v_0) when $P_t(t)$ is set to ε;

– the particle goes on sliding to the right, reaches a zero velocity, then the reaction jumps to the other side of the cone and the particle slides to the left;

– the perturbation $P_t(t)$ is then set to zero but the particle goes on sliding to the left until it reaches a zero velocity;

– the reaction then jumps to the other side of the cone and the particle starts sliding to the right, but $P_t(t)$ is then set to ε and so on.

The phase difference is qualitatively represented in Figure 5.9.

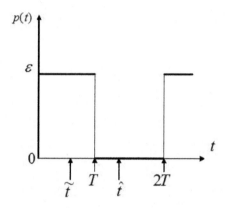

Figure 5.9. *The phase difference on the left of zone* Ω_1

Such a trajectory is the solution of the following system:

$$
\begin{cases}
\ddot{u}_1 + \omega_\alpha^2 u_1 = F_t - \mu F_n + \varepsilon, \ t \in (0, \tilde{t}) \\
u_1(0) = u_0, \ \dot{u}_1(0) = v_0, \\
\tilde{t} \text{ such that } \dot{u}_1(\tilde{t}) = 0, \\[4pt]
\ddot{u}_2 + \omega_\beta^2 u_2 = F_t + \mu F_n + \varepsilon, \ t \in (\tilde{t}, T) \\
u_2(\tilde{t}) = u_1(\tilde{t}), \ \dot{u}_2(\tilde{t}) = 0, \\[4pt]
\ddot{u}_3 + \omega_\beta^2 u_3 = F_t + \mu F_n, \ t \in (T, \hat{t}) \\
u_3(T) = u_2(T), \ \dot{u}_3(T) = \dot{u}_2(T), \\
\hat{t} \text{ such that } \dot{u}_3(\hat{t}) = 0, \\[4pt]
\ddot{u}_4 + \omega_\alpha^2 u_4 = F_t - \mu F_n, \ t \in (\hat{t}, 2T) \\
u_4(\hat{t}) = u_3(\hat{t}), \ \dot{u}_4(\hat{t}) = 0,
\end{cases}
\qquad [5.14]
$$

and the existence of a periodic solution of period $2T$ is obtained by computing a solution to the following system:

$$
u_4(2T) = u_0, \ \dot{u}_4(2T) = v_0. \qquad [5.15]
$$

From system [5.14], the periodicity condition [5.15] can be rewritten as an algebraic equation involving the unknowns u_0 and v_0:

$$
\begin{cases}
A_4 \omega_\alpha \sin \omega_\alpha (2T - \hat{t}) = -v_0, \\[4pt]
A_4 \cos \omega_\alpha (2T - \hat{t}) + \dfrac{F_t - \mu F_n}{\omega_\alpha^2} = u_0 \\[6pt]
\text{with} \\[4pt]
A_4 = \sqrt{A_3^2 + B_3^2} + \dfrac{F_t + \mu F_n}{\omega_\beta^2} - \dfrac{F_t - \mu F_n}{\omega_\alpha^2} \\[6pt]
\hat{t} = T + \dfrac{1}{\omega_\beta} \arctan \dfrac{B_3}{A_3} \\[6pt]
\text{with} \\[4pt]
A_3 = A_2 \cos \omega_\beta (T - \tilde{t}) + \dfrac{\varepsilon}{\omega_\beta^2}, \\[6pt]
B_3 = -A_2 \sin \omega_\beta (T - \tilde{t}), \\[6pt]
\text{with} \\[4pt]
A_2 = \sqrt{A_1^2 + B_1^2} + \dfrac{F_t - \mu F_n + \varepsilon}{\omega_\alpha^2} - \dfrac{F_t + \mu F_n + \varepsilon}{\omega_\beta^2} \\[6pt]
\tilde{t} = \dfrac{1}{\omega_\alpha} \arctan \dfrac{B_1}{A_1} \\[6pt]
\text{with} \\[4pt]
A_1 = u_0 - \dfrac{F_t - \mu F_n + \varepsilon}{\omega_\alpha^2} \\[6pt]
B_1 = \dfrac{1}{\omega_\alpha} v_0.
\end{cases}
\qquad [5.16]
$$

This intricate algebraic system of two equations with unknowns (u_0, v_0) is solved using the Maple software. For the values given in equation [5.5], the values (u_0, v_0) are computed and the periodic solution satisfying system [5.14] can be plotted (see Figure 5.8(b)).

5.2.1.2.3. Boundary Γ_{23} : where periodic solutions of zone Ω_2 no longer exist

The existence of a solution to problem [5.16] proves the existence of periodic solutions on the left of zone Ω_1. But these solutions exist only for T sufficiently close to the value T_0 given by [5.13]. As a matter of fact, the solution to problem [5.14]–[5.16] for different values of T shows that the amplitude of the periodic solution increases progressively as T decreases in such a way that the maximum value u_{max} of the sliding tangential displacement is reached for a reaction closer and closer to the vertex of the cone. If the reaction of a solution of problem [5.14]–[5.16] goes through the vertex, then the solution does not fulfill the unilateral contact condition and is therefore no longer a solution to the initial sliding problem. The boundary for this occurrence is determined in the following way:

i) the solution to system [5.14] is such that $u_{max} = u_1(\tilde{t})$;

ii) from problem [5.1], the vertex of the cone, that is $R_n = 0$, corresponds to a tangential displacement equal to $\dfrac{F_n}{W}$;

iii) a linear relation between u_t and R_n, $W u_t = F_n + R_n$, is deduced from problem [5.1] for $u_n \equiv 0$;

iv) consequently, the algebraic system [5.16] can be seen as a map that associates the maximal value of R_n to any pair (T, ε) for which there exists a periodic solution of the type [5.14]–[5.16]. Let $\mathcal{H}(T, \varepsilon)$ be this map.

The occurrence of solutions losing contact in the $\{T, \varepsilon\}$ plane is then given by the implicit solution $\varepsilon = \varepsilon(T)$ of the following equation:

$$\mathcal{H}(T, \varepsilon) = 0, \tag{5.17}$$

so that equation [5.17] defines the boundary Γ_{23}.

5.2.1.2.4. Zone $\Omega_5 : T_\beta < T < T_\alpha$

For decreasing values of T, the solution to equation [5.17] decreases down to values of ε very close to the limit of the range where there exist only equilibrium solutions. This curve is nevertheless not tangent to the horizontal limit of stationary solutions. There remains a thin layer in which periodic solutions with a new kind of orbit will be found. As suggested by numerical experiments, periodic solutions in this layer satisfy the following system:

$$\begin{cases} \ddot{u}_1 + \omega_\alpha^2 u_1 = F_t - \mu F_n + \varepsilon, \ t \in (0, T) \\ u_1(0) = u_0, \ \dot{u}_1(0) = 0, \\[6pt] \ddot{u}_2 + \omega_\alpha^2 u_2 = F_t - \mu F_n, \ t \in (T, \tilde{t}) \\ u_2(T) = u_1(T), \ \dot{u}_2(T) = \dot{u}_1(T), \\ \tilde{t} \text{ such that } \dot{u}_2(\tilde{t}) = 0, \\[6pt] \ddot{u}_3 + \omega_\beta^2 u_3 = F_t + \mu F_n, \ t \in (\tilde{t}, \hat{t}) \\ u_3(\tilde{t}) = u_2(\tilde{t}), \ \dot{u}_3(\tilde{t}) = 0, \\ \hat{t} \text{ such that } \dot{u}_3(\hat{t}) = 0, \\[6pt] \hat{t} \leq 2T, \ \dot{u}_4(t) = 0, \ t \in (\hat{t}, 2T) \\ u_4(2T) = u_0. \end{cases} \tag{5.18}$$

The periodicity condition reduces to $u_3(\hat{t}) = u_0$, which leads to an algebraic system much simpler than [5.16] for the determination of the initial data u_0. The solution is represented in Figure 5.10. It is clear from the conditions written in problem

[5.18] that such solutions exist only if $T_\beta < T < T_\alpha$. Moreover, the initial data of this kind of solution are such that u_0 is strictly positive by construction. It is then natural to look for the boundary of zone Ω_5 by calculating the loss of positivity of the initial data in the range $T_\beta < T < T_\alpha$. This could be done in the same way as for the boundary Γ_{23} by studying the implicit solution of equation [5.17]. But it is easily calculated that in Ω_5 the maximum u_{max} of the sliding displacement never corresponds to a reaction at the vertex of the cone, so that, due to the continuity of the solution with respect to T and ε there necessarily remains a very thin layer between Ω_5 and Ω_3.

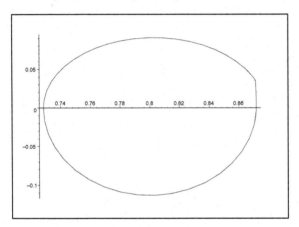

Figure 5.10. *A periodic solution in Ω_5*

5.2.1.2.5. Zone Ω_4 : periodic solutions for small T

The procedure [5.14] can easily be adapted to prove that periodic solutions exist even for very high frequencies of the excitation. The solutions in zones Ω_2 and Ω_5 suggest that zone Ω_4 may contain periodic solutions starting with a negative velocity. This means that system [5.14] should be changed into:

$$\begin{cases} \ddot{u}_1 + \omega_\beta^2 u_1 = F_t + \mu F_n + \varepsilon, \ t \in (0, \tilde{t}) \\ u_1(0) = u_0, \ \dot{u}_1(0) = v_0, \\ \tilde{t} \text{ such that } \dot{u}_1(\tilde{t}) = 0, \\[4pt] \ddot{u}_2 + \omega_\alpha^2 u_2 = F_t - \mu F_n + \varepsilon, \ t \in (\tilde{t}, T) \\ u_2(\tilde{t}) = u_1(\tilde{t}), \ \dot{u}_2(\tilde{t}) = 0, \\[4pt] \ddot{u}_3 + \omega_\alpha^2 u_3 = F_t - \mu F_n, \ t \in (T, \hat{t}) \\ u_3(T) = u_2(T), \ \dot{u}_3(T) = \dot{u}_2(T), \\ \hat{t} \text{ such that } \dot{u}_3(\hat{t}) = 0, \\[4pt] \ddot{u}_4 + \omega_\beta^2 u_4 = F_t + \mu F_n, \ t \in (\hat{t}, 2T) \\ u_4(\hat{t}) = u_3(\hat{t}), \ \dot{u}_4(\hat{t}) = 0. \end{cases} \qquad [5.19]$$

System [5.19] corresponds to trajectories represented in Figure 5.11 in the $\{R_t, R_n\}$ plane. The periodicity condition is exactly the same as equation [5.15], with an explicit form very close to [5.16]. Initial data (u_0, v_0) is thus obtained and system [5.19] gives the corresponding periodic solution in the same way as seen previously.

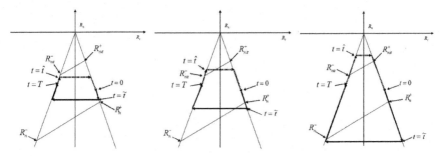

Figure 5.11. *Periodic solutions for small values of T in the $\{R_t, R_n\}$ plane*

In fact, system [5.19] does not require T to be very small, so that for the same value of ε chosen sufficiently large, the half-period T is increased progressively and periodic solutions of increasing amplitudes are obtained. Figure 5.12 contains two such periodic solutions represented at the same scale so the increase in the amplitude of the periodic solution obtained by increasing the half-period by 50% is very visible. The corresponding values of the initial data given in the caption are computed through the above scheme. Figure 5.12 brings us to formulate the following remark.

REMARK 5.3.–

i) these periodic solutions are all homeomorphic to an ellipse;

ii) for a given ε, the amplitude of the sliding oscillation decreases as T decreases;

iii) for a given T, the amplitude of the sliding oscillation decreases as ε increases;

iv) the phase difference increases as T decreases. Moreover, the phase difference is always larger in zone Ω_4 than in zone Ω_2.

5.2.2. The transition for losing contact

The $\{T, \varepsilon\}$ plane has been partitioned into zones where different kinds of periodic solutions exist. These solutions strictly satisfy the Coulomb friction law but they are bilateral by which it is meant that they satisfy the unilateral contact conditions $U_n \leq 0$, $R_n \leq 0$, $U_n R_n = 0$ by satisfying $U_n = 0$, $R_n < 0$. The horizontal line $\varepsilon = \frac{2\mu A}{K_t + \mu W}$ for any T corresponds to the lower limit of the existence of periodic solutions, which means in particular that this line is the lower boundary of all the zones of existence of any type of periodic solution described up to now. Concerning the upper boundary of these zones, the boundary Γ_{23}, transition from zone Ω_2 to zone Ω_3 where periodic solutions lose contact, has already been determined. In this section, this upper boundary, defined by the occurrence of one point losing contact during a period, is extended to the whole T axis.

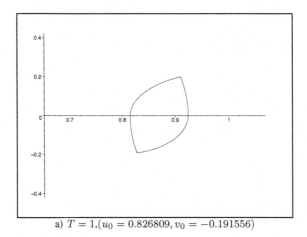

a) $T = 1, (u_0 = 0.826809, v_0 = -0.191556)$

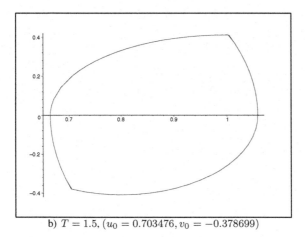

b) $T = 1.5, (u_0 = 0.703476, v_0 = -0.378699)$

Figure 5.12. *Periodic solutions in the phase space* $(u_t, \dot{u}_t)$ *when* $\varepsilon = 1.5$ *and* (u_0, v_0) *are obtained by [5.19]*

5.2.2.1. *Boundary* Γ_{34} : *from very small periods to* T_β

Let $\widetilde{\mathcal{H}}(T, \varepsilon)$ be the maximal value of the normal component of the reaction corresponding to a sliding periodic solution for a given (T, ε). Then at the transition for losing contact, T and ε satisfy:

$$\widetilde{\mathcal{H}}(T, \varepsilon) = 0. \tag{5.20}$$

In section 5.2.1.2.3, the mapping $\widetilde{\mathcal{H}}$ was defined by equation [5.17] built from problems [5.14]–[5.16]. Starting here from problem [5.19], the transition for losing contact is given in the $\{T, \varepsilon\}$ plane by the implicit solution of equation [5.20] from T close to zero to a neighborhood of T_β.

5.2.2.2. *From* T_β *to* T_α

Zone Ω_5 of Figure 5.6 has been defined for $T \in]T_\beta$, $T_\alpha[$. In this range, there exists a periodic solution of the type represented in Figure 5.10. However, everywhere in zone Ω_5, these periodic solutions are such that:

$$\widetilde{\mathcal{H}}(T, \varepsilon) < 0, \tag{5.21}$$

which means that the periodic solutions in zone Ω_5 never reach the transition for losing contact. This implies that there must exist a transition zone between zone Ω_5 and zone Ω_3 where another type of solution would lose contact as ε increases, at least for T in some sub-interval of $]T_\beta$, $T_\alpha[$. A thin layer of periodic solutions can in fact be found by solving the following system [5.22] obtained by extending zone Ω_5 after the loss of positivity of the initial position[1].

$$\begin{cases} \ddot{u}_1 + \omega_\alpha^2 u_1 = F_t - \mu F_n + \varepsilon, \ t \in (0, T) \\ u_1(0) = u_0, \ \dot{u}_1(0) = v_0, \\ \ddot{u}_2 + \omega_\alpha^2 u_2 = F_t - \mu F_n, \ t \in (T, \tilde{t}) \\ u_2(T) = u_1(T), \ \dot{u}_2(T) = \dot{u}_1(T), \\ \tilde{t} \text{ such that } \dot{u}_2(\tilde{t}) = 0, \\ \ddot{u}_3 + \omega_\beta^2 u_3 = F_t + \mu F_n, \ t \in (\tilde{t}, \tilde{t} + T_\beta), \\ u_3(\tilde{t}) = u_2(\tilde{t}), \ \dot{u}_3(\tilde{t}) = 0, \\ \ddot{u}_4 + \omega_\alpha^2 u_4 = F_t - \mu F_n, \ t \in (\tilde{t} + T_\beta, 2T) \\ u_4(\tilde{t} + T_\beta) = u_3(\tilde{t} + T_\beta), \ \dot{u}_4(\tilde{t} + T_\beta) = 0, \end{cases} \tag{5.22}$$

---

1 Let us compare problems [5.18] and [5.22]. It appears that proving that [5.18] has a solution everywhere in zone Ω_5 amounts to proving that problem [5.22] has a solution such that u_4 is constant everywhere in a non-zero measure subset of the $\{T, \varepsilon\}$-plane, which could have been missed by a direct study of problem [5.22], and which is a result interesting in itself. Moreover, distinguishing between zones Ω_5 and Ω_6 seems easier for an intuitive introduction to the partition of the plane. Nevertheless, the distinction between these two zones would not be necessary if we were only dealing with the transition to the loss of contact.

together with the periodicity condition:

$$u_4(2T) = u_0, \quad \dot{u}_4(2T) = v_0. \tag{5.23}$$

Let R_{max} be the maximal value of the normal component of the reaction. Then, using the solution of problem [5.22] by [5.23], the equality $R_{max} = 0$ is reached for some values of T and ε, which are again obtained as the implicit solution of equation [5.20]. This gives zone Ω_6 of Figure 5.6 and the periodic solutions are represented in Figure 5.13.

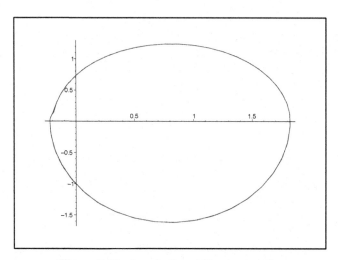

Figure 5.13. *A periodic solution in zone Ω_6*

5.2.2.3. *From T_α to $T_\alpha + T_\beta$*

The transition between zone Ω_1 and zone Ω_2 was explored in section 6.2.1.2 when studying zone Ω_1. Since the periodic solutions in zone Ω_1 are given by an explicit formula, the function that gives the boundary Γ_{12} is expressed analytically[2], and the domain of this function on the T axis appears to be divided into two sub-intervals according to whether the first loss of validity of the periodic solutions in zone Ω_1 is a loss of contact or a phase difference. In the left sub-interval, the transition already described by the occurrence of a phase difference was found, but for larger values of T and in a range corresponding to large values of ε, the loss of contact occurs before the phase difference occurs. In other words, this means that when the dynamical problem is given by equations [5.11]–[5.12], the solution $\varepsilon = \varepsilon(T)$ of equation [5.20], which

---

2 In fact, in this range the complete calculations are logically also carried out using Maple, but through an explicit formula.

is easier to compute here, because it is more explicit than in the general case, is sought for only in a sub-interval of $]T_\alpha, T_\alpha + T_\beta[$.

5.2.2.4. From $T_\alpha + T_\beta$ to very large periods

Proposition [5.1] in section 5.2.1.1 gave the existence of infinitely many periodic solutions everywhere in the open range $]T_\alpha + T_\beta, +\infty[\times]\frac{2\mu A}{K_t + \mu W}, \frac{4\mu A}{K_t + 3\mu W}[$ and a single periodic solution on the upper boundary of this range, which is the line $]T_\alpha + T_\beta, +\infty[\times \{\frac{4\mu A}{K_t + 3\mu W}\}$, and condition [5.21] is easily found to be satisfied all along this line. This means that none of the periodic solutions given by proposition 5.1 are on the point of losing contact. Again, this suggests that there might exist another zone where another type of sliding periodic solution exists up to the loss of contact[3]. Since the upper boundary $\varepsilon = \frac{4\mu A}{K_t + 3\mu W}$ is exactly the coalescence of the two loops of periodic solutions of period $4T$ into the single loop of a solution of period $2T$, these new periodic solutions may again have two loops. If so, such solutions would satisfy the following system:

$$
\begin{cases}
\ddot{u}_1 + \omega_\alpha^2 u_1 = F_t + \varepsilon - \mu F_n + \varepsilon, \quad t \in (0, T_\alpha) \\
u_1(0) = u_0, \quad \dot{u}_1(0) = 0, \\[2mm]
\ddot{u}_2 + \omega_\beta^2 u_2 = F_t + \varepsilon + \mu F_n, \quad t \in (T_\alpha, T_\alpha + T_\beta) \\
u_2(T_\alpha) = u_1(T_\alpha), \quad \dot{u}_2(T_\alpha) = 0, \\[2mm]
\ddot{u}_3 + \omega_\alpha^2 u_3 = F_t + \varepsilon - \mu F_n, \quad t \in (T_\alpha + T_\beta, T) \\
u_3(T_\alpha + T_\beta) = u_2(T_\alpha + T_\beta), \quad \dot{u}_3(T_\alpha + T_\beta) = 0, \\[2mm]
\ddot{u}_4 + \omega_\alpha^2 u_4 = F_t - \mu F_n, \quad t \in (T, \tilde{t}) \\
u_4(T) = u_3(T), \quad \dot{u}_4(T) = \dot{u}_3(T), \\
\tilde{t} \text{ such that } \dot{u}_4(\tilde{t}) = 0, \\[2mm]
\ddot{u}_5 + \omega_\beta^2 u_5 = F_t + \mu F_n, \quad t \in (\tilde{t}, \tilde{t} + T_\beta) \\
u_5(\tilde{t}) = u_4(\tilde{t}), \quad \dot{u}_5(\tilde{t}) = 0, \\[2mm]
u_6(t) = u_5(\tilde{t} + T_\beta), \quad \dot{u}_6(t) = 0, \quad t \in (\tilde{t} + T_\beta, 2T).
\end{cases}
\qquad [5.24]
$$

The periodicity condition is now $u_5(\tilde{t} + T_\beta) = u_0$, which again leads to an algebraic system of the same type as [5.16] for the determination of the initial data u_0. Periodic solutions in this range have orbits of the type represented in Figure 5.14,

---

3 A difficulty has already been encountered when studying the transition from zone Ω_5 to zone Ω_6: since little is known about the qualitative behavior, another guess could be that the loss of contact arises only through non-periodic solutions so that the line $]T_\alpha + T_\beta, +\infty[\times \{\frac{4\mu A}{K_t + 3\mu W}\}$ would be the boundary for the loss of periodicity, instead of a transition between different types of periodic solutions. The answer is given if problem [5.24] has a solution.

and the condition for losing contact is obtained as was the case before through the implicit solution of equation [5.20].

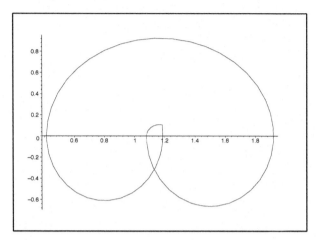

Figure 5.14. *Sliding periodic solution for large ε and $T > T_\alpha + T_\beta$*

5.2.3. *Loss of contact*

When trajectories lose contact, the calculations become more intricate. As proved by elementary examples (see Chapter 3), a trajectory may involve impacts which accumulate at some time even under extremely smooth or constant loading. But even in smoother cases where no accumulation of impacts occurs, a second-order system of dimension 2 has to be solved containing six unknowns (u_t, $\dot{u}_t$, u_n, $\dot{u}_n$, R_t and R_n), so to close the problem both the unilateral contact conditions and the Coulomb friction law have to be added to the system. The algorithms used to solve the problem were described in Chapter 3. In the following, as no accumulation of impacts is detected, an event-driven method is chosen. As a result, the tangential velocity after an impact is obtained by determining the tangential velocity and the tangential component of the reaction that satisfy the differential system and also satisfy the Coulomb law. As was shown in Chapter 3, this turns out to be the simple determination of the intersection of a straight line with the graph of the Coulomb law. An example of a trajectory involving loss of contact is shown in Figure 5.15. The restitution coefficient e is taken equal to zero, so the normal component of the velocity to the right of the impact time is zero. The period and amplitude of the excitation have been chosen inside the domain Ω_3 so loss of contact is expected at some time during the period.

Figure 5.15 represents a solution to system [5.1] with a rectangular wave shape perturbation as before. The period of the excitation is equal to 7, the same period as for the periodic solutions shown in Figure 5.5, but its amplitude is strictly larger than the limit $\varepsilon(T)$ where contact is lost. The solution remains periodic of period $4T$. The normal component of the displacement does not stay identically equal to zero as shown in Figure 5.15(b) and satisfies the unilateral contact condition. Concerning the tangential component of the displacement, Figure 5.15(a) shows that the period of the response is twice the one of the excitation as was the case for $\varepsilon < \varepsilon(T)$ for the same value of T.

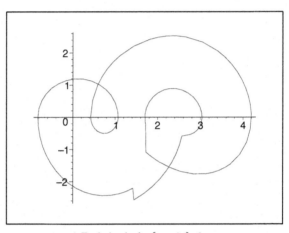

a) Evolution in the $\{u_t, \dot{u}_t\}$ plane

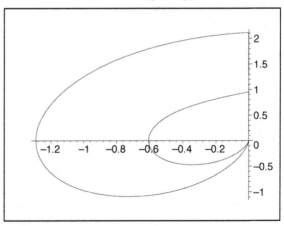

b) Evolution in the $\{u_n, \dot{u}_n\}$ plane

Figure 5.15. *Loss of contact*

The time intervals where u_n is strictly negative, that is during which contact is lost, end by an impact associated with a jump of $\dot{u}_t$. Such impacts occur twice during the period (the normal velocity is not zero to the left of these impact times as shown in Figure 5.15(b)) but Figure 5.15(a) shows that at one of the impacts the tangential velocity simply reduces its amplitude, whereas at the other impact the velocity jumps from a strictly negative value to zero.

5.2.4. *The global behavior*

The qualitative behavior of the periodic solutions of problem [5.1] submitted to a rectangular wave shape perturbation is summarized in Figure 5.16[4]. The following comments can be formulated, paying particular attention to the continuous curve, represented by $\varepsilon(T)$, which corresponds to the loss of contact:

– the minimum value of $\varepsilon(T)$ is very close to the loss of existence of equilibrium states. This means that this minimum could be interpreted as a resonance;

– conversely, when $T \longrightarrow 0$ there exist periodic solutions strictly in contact for values of ε, which are large compared to the largest value of ε for which there exists an equilibrium under static loading. Using the values given in [5.5], the latter is $\varepsilon = 3$ while $\varepsilon(T) \simeq 6$ for T close to zero. The curve $\varepsilon(T)$ is far from reaching this value in any other zone;

– the intricate behavior of the periodic solutions is localized in a very small and thin range close to the minimum of $\varepsilon(T)$;

– there is only one zone where there exists more than one periodic solution for each given value of T and ε. This zone is a horizontal strip that goes from $T_\alpha + T_\beta$ to $+\infty$. Everywhere in this strip, there exist infinitely many periodic solutions. All these solutions are such that R_n is strictly negative at any time so that they all remain strictly in contact.

5.3. When a single equilibrium solution exists

This section is concerned with the transition between the zone where periodic solutions exist ($\varepsilon > \varepsilon_0$) and the zone where there exist infinitely many equilibrium solutions ($\varepsilon < \varepsilon_0 = \dfrac{2\mu A}{K_t + \mu W}$). For this specific value ε_0 of the amplitude, there exists a unique equilibrium solution for any value of the half-period T (see theorem 5.1). However, it is shown in this section that if the period of the external force is large enough, there also exist infinitely many periodic solutions in addition to the single equilibrium state. The stability of this unique equilibrium state depends on the value of the period of the excitation. This equilibrium state will be shown to be asymptotically stable for $T < T_\alpha$ (i.e. all trajectories will converge to the equilibrium state) or Lyapunov stable for $T > T_\alpha$.

---

4 It is interesting to observe that, while giving the transition between zone Ω_1 and zone Ω_2 only amounts to plotting the graph of an explicit function, calculating all the other curves of Figure 5.16 requires an implicit computation that can be very time consuming.

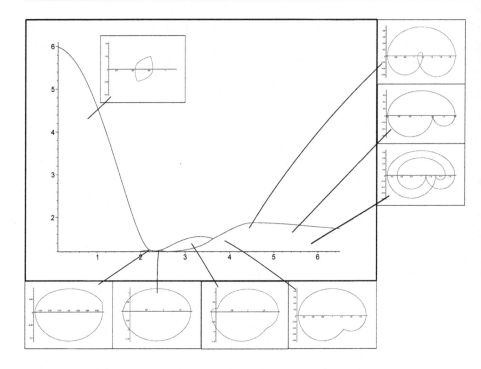

Figure 5.16. *The global behavior in the period–amplitude plane*

REMARK 5.4.– It is interesting to note that:

i) The equilibrium state is reached in general at infinity.

ii) The amplitude of the oscillations takes more and more time to decrease as T increases toward T_α, and beyond T_α these oscillations change into periodic solutions. This suggests that the point $(T, \varepsilon) = (T_\alpha, \frac{2\mu A}{K_t + \mu W})$ of the $\{period, amplitude\}$ plane could be interpreted as a Hopf bifurcation point.

iii) Such a Hopf-type bifurcation point is nevertheless not classical since all the periodic solutions lie in a bounded domain diffeomorphic to an ellipse and are surrounded by non-periodic solutions similar to those obtained before the bifurcation point but converging this time toward a periodic solution. This is represented in Figure 5.17.

Using the numerical values given in [5.5], this equilibrium state is the point $(u_t = 0.8, \dot{u}_t = 0)$ of the phase plane.

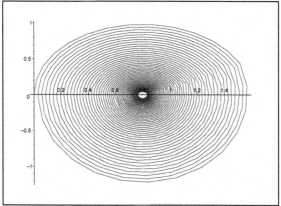

a) Convergence toward the single stationarysolution for $T < T_\alpha$

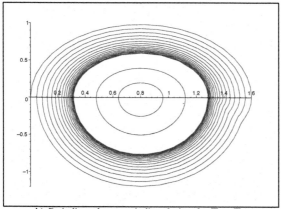

b) Periodic and non-periodic solutionsfor $T > T_\alpha$

Figure 5.17. *Solutions for $\varepsilon = 1.2$*

The following proposition will be proved in the sections that follows.

PROPOSITION 5.2.– For $\varepsilon = \dfrac{2\mu A}{K_t + \mu W}$, a single equilibrium solution exists whatever the value of the half-period T. Moreover:

i) when $T < T_\alpha$, all the initial data different from the single equilibrium lead to trajectories that tend to the equilibrium;

ii) when $T > T_\alpha$, there exist infinitely many periodic solutions and any initial data that do not belong to a periodic solution lead to a trajectory that tends to a periodic solution;

iii) when $T = T_\alpha$, the set of periodic solutions is strictly larger than the set of periodic solutions obtained for $T > T_\alpha$. Any initial data that do not belong to a periodic solution lead to a trajectory that tends to the periodic solution of largest orbit at infinity.

5.3.1. *The set of solutions when T is larger than T_α*

5.3.1.1. *When T is strictly larger than T_α*

When the amplitude ε of the perturbation is equal to $\varepsilon_0 = \frac{2\mu A}{K_t + \mu W}$, there exists a single equilibrium solution for any value of the half-period T, let U_e be this equilibrium position. Other specific positions will prove to be useful in the following analysis:

– U_ℓ the equilibrium position in imminent sliding to the left when the perturbation is equal to ε_0;

– U_r the equilibrium position in imminent sliding to the right with the perturbation equal to 0;

– U_d the position such that all trajectories issued from $(u_0, 0)$ with $u_0 < U_d$ lose contact.

Using the fact that a trajectory will no longer be a sliding trajectory (i.e. the mass loses contact) as soon as its reaction goes through the vertex of the cone, it is easy to establish that the positions introduced above are given by the following expressions:

$$
\begin{cases}
U_e = \dfrac{F_n}{W} + \dfrac{-A + \varepsilon_0 W}{W\omega_\alpha^2} = \dfrac{F_n}{W} - \dfrac{A}{W\omega_\beta^2} \\
\text{with normal reaction } R_e = \dfrac{-A + \varepsilon_0 W}{\omega_\alpha^2}, \\[2ex]
U_d = \dfrac{F_n}{W} - \dfrac{2A}{W\omega_\beta^2} \\
\text{with normal reaction } R_d = \dfrac{-2A}{\omega_\beta^2}, \\[2ex]
U_\ell = \dfrac{F_n}{W} + \dfrac{-A + \varepsilon_0 W}{W\omega_\beta^2} \equiv \dfrac{F_t + \mu F_n + \varepsilon_0}{\omega_\beta^2} \\
\text{with normal reaction } R_\ell = \dfrac{-A + \varepsilon_0 W}{\omega_\beta^2}, \\[2ex]
U_r = \dfrac{F_n}{W} - \dfrac{A}{W\omega_\alpha^2} \equiv \dfrac{F_t - \mu F_n}{\omega_\alpha^2} \\
\text{with normal reaction } R_r = \dfrac{-A}{\omega_\alpha^2}.
\end{cases}
\tag{5.25}
$$

Introducing the quantity d defined as

$$d = U_\ell - U_e = \frac{F_n}{W} + \frac{-A + \varepsilon_0 W}{W \omega_\beta^2} - U_e = \frac{\varepsilon_0}{\omega_\beta^2} > 0,$$

so the symmetric of U_ℓ with respect to U_e will be simply $U_e - d$.

REMARK 5.5.–

i) The numerical values chosen in [5.5] give $U_r = 0$.

ii) The inequalities $U_r < U_e - d < U_e$ and $U_d < U_e - d$ always hold. In particular, inequality $U_d < U_e - d$ results from the condition $K_t - \mu W > 0$, which has been referred to as the generic case and which is consequently assumed to hold all through this chapter.

iii) It is easy to check that $U_d < U_r \iff K_t - 3\mu W > 0$. Therefore, the value of U_d depends on the mechanical parameters of the system.

Using the notations introduced in [5.25], when the tangential external load $F_t(t) = F_t + \varepsilon$, that is when $t \in]2iT, (2i+1)T]$, the set of normal components of the equilibria is given by $\{R_n\}(t) = [R_e, R_\ell]$, and when the tangential external load $F_t(t) = F_t$, that is when $t \in](2i+1)T, (2i+2)T]$, the set of normal components of the equilibria is given by $\{R_n\}(t) = [R_r, R_e]$ (see Figure 5.18(a)).

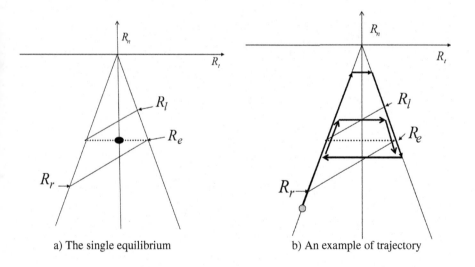

a) The single equilibrium   b) An example of trajectory

Figure 5.18. *The single equilibrium for $\varepsilon = \varepsilon_0$ in the (R_t, R_n) plane*

As was the case beforehand, the notation u is used in place of u_t (as long as only sliding motions are considered no confusion appears). Let the external loading be such that $\varepsilon = \varepsilon_0$. Then, the qualitative behavior of the solutions to system [5.1] with $u_n \equiv 0$ is given by propositions 5.3 and 5.4.

PROPOSITION 5.3.– Let $T > T_\alpha$, $\varepsilon = \varepsilon_0$ and $K_t - \mu W > 0$, then any initial data $(u_0, \dot{u}_0)$ such that $u_0 \in [U_e - d, U_e]$ and $\dot{u}_0 = 0$ lead to a periodic solution of period $2T$.

PROOF.– The load changes periodically from (F_t, F_n) to $(F_t + \varepsilon_0, F_n)$ with a period $2T$. It is equal to $(F_t + \varepsilon_0, F_n)$ on $[0, T[$ and to (F_t, F_n) on $[T, 2T[$. If $u_0 \in [U_e - d, U_e]$, then $R_n(T_\alpha)$ belongs to $[R_e, R_\ell]$ and lemma 5.1 implies that the mass remains motionless for $t \in [T_\alpha, T]$. When the load changes, the mass starts sliding to the left and once again lemma 5.1 implies that the mass will remain motionless for $t \in [T + T_\beta, 2T]$. For $u_0 \in [U_e - d, U_e]$, the motion is then governed by the following system:

$$\begin{cases} \ddot{u} + \omega_\alpha^2 u = F_t + \varepsilon - \mu F_n, & t \in [0, T_\alpha], \\ u(0) = u_0, \ \dot{u}(0) = 0, \\ u(t) = u(T_\alpha), \ \dot{u}(t) = 0, & t \in [T_\alpha, T], \\ \ddot{u} + \omega_\beta^2 u = F_t + \mu F_n, & t \in [T, T + T_\beta], \\ u(T) = u(T_\alpha), \ \dot{u}(T) = 0, \\ u(t) = u(T + T_\beta), \ \dot{u}(t) = 0, & t \in [T + T_\beta, 2T]. \end{cases} \qquad [5.26]$$

A solution to system [5.26] is such that:

$$u(T) = u(T_\alpha) = U_e + (U_e - u_0)$$

and

$$u(2T) = u(T + T_\beta) = U_e - (u(T) - U_e) = u_0, \dot{u}(2T) = 0.$$

In other words, the solution is $2T$ periodic. □

REMARK 5.6.–

i) In all the figures, the use of the numerical values defined at equation [5.5] implies that the interval $[U_e - d, U_e + d]$ is equal to $[0.32, 1.28]$;

ii) Studying only the case of initial data with zero velocity may seem restrictive. In fact, introducing a non-zero initial velocity would not lead to qualitative changes;

iii) Choosing $u_0 < U_e$ implies that the first phase of the motion is a sliding phase to the right. If $u_0 \geq U_e$, then the first phase would be either motionless or sliding to the left but no other periodic solutions would be found.

Of course, if $u_0 < U_e - d$, then $R_n(T_\alpha)$ is greater than R_ℓ, therefore $R_n(T_\alpha) \notin [R_e, R_\ell]$ so the mass does not stay motionless in the time interval $[T_\alpha, T]$ and the motion is no longer governed by system [5.26]. Whether in the $\{R_t, R_n\}$ plane or in the phase space $\{u, \dot{u}\}$, the set of periodic trajectories completely fills the inside of the domain bounded by the periodic trajectory of largest amplitude represented in Figure 5.19. Proposition 5.4 will prove how trajectories starting out of interval $[U_e - d, U_e + d]$ behave and will also show that there are no other periodic solutions than those given by proposition 5.3. The first step consists of showing that all sliding trajectories enter the interval $[U_r, U_e - d[$ in finite time.

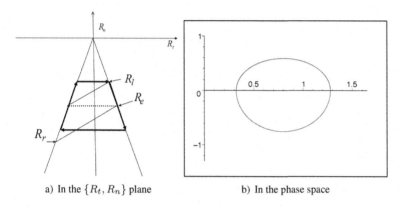

a) In the $\{R_t, R_n\}$ plane   b) In the phase space

Figure 5.19. *The periodic solution of maximal amplitude for $\varepsilon = \varepsilon_0$ and $T > T_\alpha$*

LEMMA 5.2.– Let $K_t - 3\mu W > 0$, $T > T_\alpha$ and assume that the initial position belongs to the interval $[U_d, U_r[$ with an initial velocity equal to 0. Then, the trajectory intersects the interval $[U_r, U_e - d[$ in finite time.

PROOF.– As the period of the excitation is strictly greater than T_α, it is written as $T = T_\alpha + \eta$ with some strictly positive η. Assume that after a trajectory involving k non-periodic loops, the mass is at some out-of-equilibrium position u_k with a zero velocity on the left side of the cone and takes this position to be the initial data at a time chosen as $t = 0$. Then, there is a sliding phase to the right, followed by a jump to the right side of the cone and by a sliding phase to the left, the trajectory then comes back to a new out-of-equilibrium position u_{k+1} on the left side of the cone, which makes a complete loop. More precisely, let $\tau_1, \tau_2, \tau_3, \tau_4$ be successive times represented in Figure 5.20 and defined as:

– during a first sliding phase to the right, τ_1 is the time when the perturbation is set to ε_0;

– τ_2 is the time when the velocity of the sliding phase to the right goes through zero, so τ_2 is the end of the sliding phase to the right;

– during the sliding phase to the left, τ_3 is the time when the perturbation is set to zero;

– τ_4 is the time when the velocity of the sliding phase to the left goes through zero again.

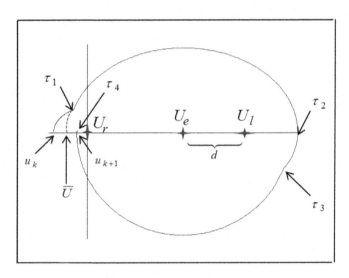

Figure 5.20. *One loop of the orbit for $T > T_\alpha$, with some particular values useful for the calculations*

Choosing $u_k \in [U_d, U_r[$ and $\dot{u}_k = 0$ as initial data, position u_{k+1} is obtained by solving the system:

$$
\begin{cases}
\ddot{u} + \omega_\alpha^2 u = F_t - \mu F_n & \text{on } (0, \tau_1) \\[2mm]
\ddot{u} + \omega_\alpha^2 u = F_t - \mu F_n + \varepsilon_0 & \text{on } (\tau_1, \tau_2) \\[2mm]
\ddot{u} + \omega_\beta^2 u = F_t + \mu F_n + \varepsilon_0 & \text{on } (\tau_2, \tau_3) \\[2mm]
\ddot{u} + \omega_\beta^2 u = F_t + \mu F_n & \text{on } (\tau_3, \tau_4),
\end{cases}
\qquad [5.27]
$$

using the continuity of the position u and of the velocity $\dot{u}$ at times τ_1, τ_2, τ_3. Position u_{k+1} is defined by $u_{k+1} := u(\tau_4)$, where τ_4 is such that $\dot{u}(\tau_4) = 0$ so that:

$$
u_{k+1} = U_e - \Big[(U_e - \overline{U} - d)^2 + d^2 \\
+ 2d(U_e - \overline{U} - d) \cos(\omega_\beta \eta) \Big]^{1/2},
$$

where $\overline{U}$ is an abstract initial position defined in the following way:

i) assume the perturbation is equal to ε_0 since the origin $t = 0$ instead of being set to ε_0 at time τ_1;

ii) then $(\overline{U}, 0)$ is the initial data that would lead to a trajectory that would coincide with the actual trajectory calculated in $[\tau_1, \tau_2]$;

iii) as represented in Figure 5.20, this means that $\overline{U}$ is symmetrical of $u(\tau_2)$ with respect to U_e.

So,

$$(U_e - u_{k+1})^2 = (U_e - \overline{U})^2 - 2d(U_e - \overline{U} - d)(1 - \cos(\omega_\beta \eta)),$$

from which

$$u_{k+1} - \overline{U} = \frac{2d(U_e - \overline{U} - d)(1 - \cos(\omega_\beta \eta))}{(U_e - \overline{U}) + (U_e - u_{k+1})}.$$

If $\overline{U} \geq U_r$, then as u_{k+1} is larger than $\overline{U}$, u_{k+1} obviously belongs to $[U_r, U_e - d[$.

If not, then

$$u_{k+1} - u_k \geq u_{k+1} - \overline{U} \geq \frac{2d(U_e - U_r - d)(1 - \cos(\omega_\beta \eta))}{(U_e - U_d) + (U_e - U_d)},$$

and setting

$$\delta = \frac{2d(U_e - U_r - d)(1 - \cos(\omega_\beta \eta))}{2(U_e - U_d)} > 0,$$

finally

$$u_{k+1} - u_k \geq \delta > 0.$$

This implies that u_k is larger than U_r as soon as $k\delta > U_r - U_d$, that is at most after $2kT$, therefore in finite time. $\square$

PROPOSITION 5.4.– Let $T > T_\alpha$, $\varepsilon = \varepsilon_0$ and $K_t - \mu W > 0$, then the trajectories issued from any initial position $u_0 \notin [U_e - d, U_e + d]$ are not periodic, and have a qualitative behavior that depends on T in the following way:

i) for any T such that $T_\alpha < T \leq T_\alpha + T_\beta/2$, all sliding solutions tend at infinity to the periodic solution of largest amplitude,

ii) for any T such that $T > T_\alpha + T_\beta/2$, all sliding solutions reach one of the periodic solutions in finite time.

PROOF.– According to the values given in equation [5.25] and in remark 5.6, proving proposition 5.4 amounts to investigating the behavior of any trajectory starting with a zero velocity and an initial position u_0 such that

$$U_d < u_0 < U_e - d.$$

Due to lemma 5.2, the proof of proposition 5.4 can be restricted to trajectories starting from initial positions in the interval $[U_r, U_e - d[$ with a zero velocity. Setting $u_k = u(2kT)$, it is easy to check that lemma 5.1 implies that for all k, u_k belongs to $[U_r, U_e - d[$, so $\tau_1 = 0$ in system [5.27] and u_{k+1} can easily be calculated.

Setting: $x_k = U_e - u_k$.

If $T \geq T_\alpha + T_\beta$, then $\dot{u}(\tau_3) = 0$ and the solution of system [5.27] gives:

$$x_{k+1} = \sqrt{d^2 + (x_k - d)^2 - 2d(x_k - d)},$$

so

$$x_{k+1} = (d - (x_k - d)) > 0.$$

But as $u_k \in [U_r, U_e - d[$, this implies that $x_k - d > 0$ so

$$x_{k+1} - d = -(x_k - d) < 0,$$

therefore, for any initial position in $[U_r, U_e - d[$, one of the periodic solutions is reached after $2T$. Such a trajectory is represented in Figure 5.21 in the $\{R_t, R_n\}$ plane.

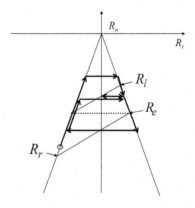

Figure 5.21. *A trajectory starting from initial positions in the interval* $[U_r, U_e - d[$ *with a zero velocity for* $T \geq T_\alpha + T_\beta$

If $T < T_\alpha + T_\beta$, the solution of system [5.27] gives:

$$x_{k+1} = \sqrt{d^2 + (x_k - d)^2 + 2d(x_k - d)\cos\omega_\beta(T - T_\alpha)}. \tag{5.28}$$

The behavior of the sequence $\{x_k\}$ defined by equation [5.28] depends on whether $\cos\omega_\beta(T - T_\alpha)$ is positive or negative so there are two cases:

i) $T_\alpha < T \le T_\alpha + T_\beta/2$: then $0 \le \cos\omega_\beta(T - T_\alpha) < 1$ and consequently

$$x_k > d \implies x_{k+1}^2 - d^2 > 0 \implies x_{k+1} > d.$$

Therefore, $\forall k$ $x_k > d$ so d is a lower bound for the sequence $\{x_k\}$. On the other hand, equation [5.28] implies that:

$$x_{k+1}^2 - x_k^2 = 2d(d - x_k)(1 - \cos\omega_\beta(T - T_\alpha)) < 0, \tag{5.29}$$

which means that the sequence $\{x_k\}$ is decreasing. Therefore, the sequence $\{x_k\}$ converges, and passing to the limit in equation [5.29], $\{x_k\}$ converges toward d, which establishes point (i) of proposition 5.4.

ii) $T_\alpha + T_\beta/2 < T < T_\alpha + T_\beta$: then $\cos\omega_\beta(T - T_\alpha) < 0$. In this case, there exists a subscript k^* such that $x_{k^*} \le d$. In other words, the trajectory converges in finite time to the periodic solution with initial condition $(u_{k^*}, 0)$.

This is established by assuming that it is not so, that is by assuming that for all k, $x_k > d$. Then, as in part (i) of the proof, equation [5.29] implies that the sequence $\{x_k\}$ converges toward d. But from equation [5.28]

$$x_{k+1}^2 - d^2 = (x_k - d)^2 + 2d(x_k - d)\cos\omega_\beta(T - T_\alpha),$$

so

$$(x_k - d)^2 + 2d(x_k - d)\cos\omega_\beta(T - T_\alpha) > 0, \quad \forall k,$$

or

$$(x_k - d) + 2d\cos\omega_\beta(T - T_\alpha) > 0, \quad \forall k.$$

But this cannot be true for all k since the quantity $2d\cos\omega_\beta(T - T_\alpha)$ is given and strictly negative, whereas x_k converges to d. So, the assumption "for all k, $x_k > d$" is false and there exists k^* such that $x_{k^*} \le d$. $\square$

Figure 5.22 represents two trajectories starting from the same initial data calculated with the numerical values given in equation [5.5]: the first one, for a large enough T, attains one of the periodic solutions in finite time, in fact after a very small number of sliding oscillations (Figure 5.22(a)), the other one, for a smaller T, converges to the largest amplitude periodic solution at infinity (Figure 5.22(b)).

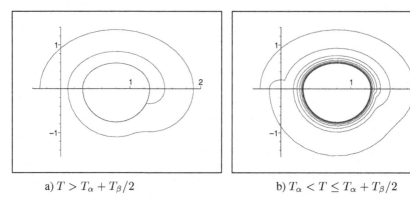

a) $T > T_\alpha + T_\beta/2$ b) $T_\alpha < T \leq T_\alpha + T_\beta/2$

Figure 5.22. *Two non-periodic trajectories for* $\varepsilon = \varepsilon_0$ *and* $T > T_\alpha$

5.3.1.2. *Periodic solutions for* $T = T_\alpha$

This section focuses on point *(iii)* of proposition 5.2.

PROPOSITION 5.5.– Assume $T = T_\alpha$ and $\varepsilon = \varepsilon_0$, then any initial data $(u_0, \dot{u}_0)$ with $u_0 \in [max(U_r, U_d), U_e]$ and $\dot{u}_0 = 0$ lead to a periodic solution of period $2T$.

PROOF.– Under the assumption $U_d \leq U_r$ (which holds for $K_t - 3\mu W \geq 0$), if u_0 is chosen in $[U_r, U_e]$, the motion will be governed by a system similar to system [5.26], except that the motionless phase during $[T_\alpha, T]$ in [5.26] does not exist because $T = T_\alpha$ so the trajectory is the solution to:

$$\begin{cases} \ddot{u} + \omega_\alpha^2 u = F_t + \varepsilon - \mu F_n, \quad t \in [0, T_\alpha], \\ u(0) = u_0, \quad \dot{u}(0) = 0, \\ \ddot{u} + \omega_\beta^2 u = F_t + \mu F_n, \quad t \in [T_\alpha, T_\alpha + T_\beta], \\ u(T_\alpha) = u(T_\alpha), \quad \dot{u}(T_\alpha) = 0, \\ u(t) = u(T_\alpha + T_\beta), \quad \dot{u}(t) = 0, \quad t \in [T_\alpha + T_\beta, 2T]. \end{cases} \qquad [5.30]$$

A solution to system [5.30] is such that:

$$\begin{cases} u(T_\alpha) = U_e + U_e - u_0, \\ u(2T) = u(T_\alpha + T_\beta) = U_e - (u(T) - U_e) = u_0, \\ \text{with } \dot{u}(2T) = 0. \end{cases}$$

In other words, the solution is $2T$ periodic.

The set of initial conditions that give a periodic solution is larger than when $T > T_\alpha$ simply because here, when $T = T_\alpha$, the normal reaction $R_n(T_\alpha)$ does not belong to $[R_e, R_\ell]$ if $u_0 \in [U_r, U_e - d[$ so there is no motionless phase. Comparing this set with the one obtained for the case $T > T_\alpha$, the set again completely fills a domain, either in the $\{R_t, R_n\}$ plane or in the phase space, which is bounded by the periodic solution of maximal amplitude, but this set is now strictly larger than for $T > T_\alpha$. It is represented in Figure 5.23, where the scales on both axis of Figures 5.23(a) and (b) are the same as those used for Figures 5.19(a) and (b). □

COROLLARY 5.1.– When $U_d > U_r$ (i.e. $K_t - 3\mu W < 0$), there exist no non-periodic sliding solutions. All the sliding trajectories are periodic, and any other trajectory involves loss of contact.

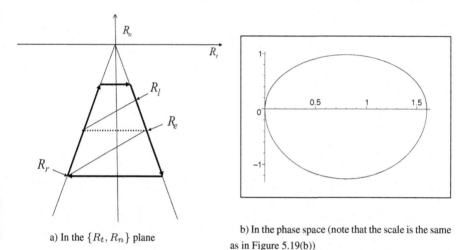

a) In the $\{R_t, R_n\}$ plane

b) In the phase space (note that the scale is the same as in Figure 5.19(b))

Figure 5.23. *The maximal amplitude periodic solution for* $\varepsilon = \varepsilon_0$ *and* $T = T_\alpha$ *with* $U_d < U_r$

5.3.2. When T is smaller than T_α

PROPOSITION 5.6.– Assume $T < T_\alpha$ and $\varepsilon = \varepsilon_0$, then any initial data such that $u_0 \in]U_d, U_e[$ and $\dot{u}_0 = 0$ lead to a trajectory that converges to the single equilibrium.

An elementary corollary of proposition 5.6 is that, due to the well-posedness of the Cauchy problem, no periodic solution exists (except the trivial one that is the single equilibrium point). The proof of proposition 5.6 is carried out in several steps, it will establish that all the trajectories of sliding motion converge to the equilibrium, but depending on the period and on the initial data the convergence holds in finite time or at infinity.

Step 1: Let $T \leq T_\alpha/2$. Then, there exists an initial data $u_0 < U_e$ such that the trajectory arrives exactly at the equilibrium point U_e with a zero velocity after a finite number of oscillations of the external force and involves only sliding phases on the left side of the cone.

To establish this existence result, the trajectory is explicitly computed. As long as the reaction remains on the left side of the cone, the position $u(t)$ is smaller than U_e and the velocity $\dot{u}(t)$ is strictly positive, the successive phases of the motion are given by:

$$
\left\{
\begin{array}{l}
\text{For } i = 0, ...n - 1, \\
\quad \text{when } t \in [2iT, (2i + 1)T[: \\
\quad u(t) = U_e + (u_0 - U_e) \cos \omega_\alpha t \\
\qquad\qquad - \dfrac{\varepsilon_0}{\omega_\alpha^2} \sum_{j=1}^{2i} (-1)^j \cos \omega_\alpha (t - jT), \\
\quad \text{when } t \in [(2i + 1)T, (2i + 2)T[: \\
\quad u(t) = U_e - \dfrac{\varepsilon_0}{\omega_\alpha^2} + (u_0 - U_e) \cos \omega_\alpha t \\
\qquad\qquad - \dfrac{\varepsilon_0}{\omega_\alpha^2} \sum_{j=1}^{2i+1} (-1)^j \cos \omega_\alpha (t - jT),
\end{array}
\right.
\qquad [5.31]
$$

where the number n of oscillations is determined through the half-period T of the load by:

$$
\frac{T_\alpha}{2n + 1} \leq T < \frac{T_\alpha}{2n - 1}.
\qquad [5.32]
$$

Consequently for any given half-period T in $]0, T_\alpha/2]$, an initial data $(u_0, 0)$ such that the trajectory leads to the equilibrium U_e with zero velocity after n oscillations can be calculated from equation [5.31]. An example of such a trajectory is represented in Figure 5.24 for $n = 7$. Obviously, a trajectory can reach U_e with a zero velocity during the interval $[0, T]$ only in the trivial case $u_0 = U_e$.

For example, in the case $T = T_\alpha/4$, that is $n = 2$ from equation [5.32], the velocity can be zero for the first time at some $\tilde{t} \in]3T, 4T]$ given by:

$$
\tan \omega_\alpha \tilde{t} = \frac{\sin \omega_\alpha T - \sin 2\omega_\alpha T + \sin 3\omega_\alpha T}{\cos \omega_\alpha T - \cos 2\omega_\alpha T + 3 \cos \omega_\alpha T + \frac{\omega_\alpha^2}{\varepsilon_0}(u_0 - U_e)}
$$

and time $\tilde{t}$ will be the final point of the trajectory if $u(\tilde{t}) = U_e$, which gives the initial data for such a trajectory:

$$\tilde{u}_0 = U_e - \frac{\varepsilon_0}{\omega_\alpha^2}\Big[\cos 2\omega_\alpha T(2\cos\omega_\alpha T - 1)$$
$$+\Big(\cos^2 2\omega_\alpha T(2\cos\omega_\alpha T - 1)^2$$
$$-4\cos\omega_\alpha T(\cos\omega_\alpha T - 1)\Big)^{1/2}\Big].$$

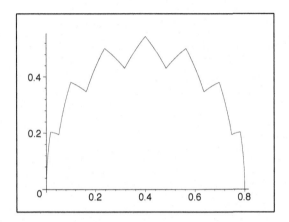

Figure 5.24. *An example of trajectory represented in the phase space converging to the single equilibrium in finite time*

From now on, $\tilde{u}_{0,\alpha}$ will denote such an initial position to stress the fact that it corresponds to sliding phases to the right (and accordingly $\tilde{u}_{0,\beta}$ will denote an equivalent initial position of sliding phases to the left, when necessary). The corresponding trajectory is represented in Figure 5.25(a).

Step 2: Let T be a given half-period of the force in $]0, T_\alpha/2]$, and $(\tilde{u}_{0,\alpha}, 0)$ the initial data leading to a trajectory that reaches the equilibrium U_e in finite time for this period. Then, any initial data $(u_0, 0)$ with $u_0 \in]\tilde{u}_{0,\alpha}, U_e[$ lead to a trajectory that converges to U_e at infinity.

The proof of Step 2 is based upon the following result:

LEMMA 5.3.– If the initial data $(u_0, 0)$ are such that $U_e - u_0$ is strictly positive and satisfies the following inequality:

$$U_e - \frac{\varepsilon_0 \cos\omega_\alpha T(1 + \tan^2\omega_\alpha T)}{2\omega_\alpha^2} < u_0, \quad \text{when } T \in]0, T_\alpha/3[\qquad [5.33]$$

$$\tilde{u}_{0,\alpha} < u_0, \quad \text{when } T \in [T_\alpha/3, T_\alpha/2] \tag{5.34}$$

then the trajectory involves only sliding phases to the right, which pile up at the left of U_e.

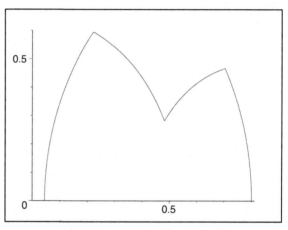

a) Initial data $(u(0), \dot{u}(0)) = (\tilde{u}_{0,\alpha}, 0)$

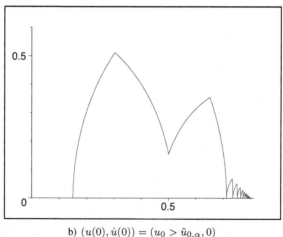

b) $(u(0), \dot{u}(0)) = (u_0 > \tilde{u}_{0,\alpha}, 0)$

Figure 5.25. *One trajectory, here represented in the phase space, converges to the single equilibrium in finite time while the other converges to the equilibrium at infinity*

PROOF.– When $T \in [T_\alpha/3, T_\alpha/2]$, the number of oscillations defined in [5.32] is equal to one and $\tilde{u}_{0,\alpha} = U_e - \dfrac{2\varepsilon_0 \cos \omega_\alpha T}{\omega_\alpha^2}$, so if $\tilde{u}_{0,\alpha} < u_0$, the position $u(t_1)$

where the velocity is zero after a single oscillation is strictly smaller than U_e. When $T \in]0, T_\alpha/3[$ inequality [5.33] implies that starting from a point u_0 at $t = 0$, the velocity will be equal to zero at some time t_1 in $]T, 2T[$. So, whatever the value of $T \le T_\alpha/2$, the motion is governed by:

$$\begin{cases} t \in [0, T]: \quad u(t) = U_e + (u_0 - U_e)\cos\omega_\alpha t, \\[2mm] t \in [T, t_1]: \quad u(t) = U_e - \dfrac{\varepsilon_0}{\omega_\alpha^2} + (u_0 - U_e)\cos\omega_\alpha t \\[1mm] \qquad\qquad\qquad +\dfrac{\varepsilon_0}{\omega_\alpha^2}\cos\omega_\alpha(t - T), \\[2mm] t \in [t_1, 2T]: \quad u(t) \equiv u(t_1). \end{cases} \qquad [5.35]$$

It is obvious that $u_1 := u(t_1) > u_0$. Then, let $u_{k-1} := u(t_{k-1})$ be a point that is at equilibrium when the perturbation is equal to zero and from which a motion starts sliding to the right when the perturbation is set to ε_0, and take $(u_{k-1}, 0)$ as initial data. Inequality [5.33] implies that the trajectory involves only equation [5.35] so that time t_k and position $u_k := u(t_k)$ when the velocity is equal to zero again are given by:

$$\begin{cases} \tan\omega_\alpha t_k = \dfrac{\sin\omega_\alpha T}{\cos\omega_\alpha T + \frac{\omega_\alpha^2}{\varepsilon_0}(u_{k-1} - U_e)}, \\[4mm] u_k = U_e - \dfrac{\varepsilon_0}{\omega_\alpha^2} + \left(\dfrac{\varepsilon_0^2}{\omega_\alpha^4} + (u_{k-1} - U_e)^2 \right. \\[2mm] \qquad\qquad \left. +\dfrac{2\varepsilon_0}{\omega_\alpha^2}(u_{k-1} - U_e)\cos\omega_\alpha T\right)^{1/2}. \end{cases} \qquad [5.36]$$

Sequence $\{u_k\}$ is thus defined from equation [5.36]. It is such that $u_{k+1} > u_k$, and $u_{k+1} < U_e$ so the sequence $\{u_k\}$ converges to a value l and by passing to the limit in the expression of u_k given in equation [5.36] this limit l is found to be equal to U_e, which completes the proof of lemma 5.3. □

Step 2 follows as a simple corollary. An example of such a case is represented in Figure 5.25(b), to be compared to 5.25(a). Indeed, convergence to the equilibrium holds at infinity, but there exist particular values of the initial data for which convergence holds in finite time; only two oscillations in the example of Figure 5.25(a).

Step 3: Let T be a given half-period of the external force, with $T \le T_\alpha/2$ and $(\tilde{u}_{0,\alpha}, 0)$ the initial data leading to a trajectory that reaches the equilibrium U_e in finite time for this period. Then, any initial data $(u_0, 0)$ with $u_0 \in]U_d, \tilde{u}_{0,\alpha}[$ lead to a trajectory that jumps to the other side of the cone and converges to U_e.

PROOF.– The special position U_d has been defined in equation [5.25] as the position such that all trajectories issued from $(u_0, 0)$ with $u_0 < U_d$ lose contact. The sketch of the proof is then as follows. First, if $u_0 \in]U_d, \tilde{u}_{0,\alpha}[$, the trajectory sliding to the right goes beyond the equilibrium U_e and, when it passes through zero, jumps to the other side of the cone. Second, as mentioned in Step 1, a position $\tilde{u}_{0,\beta}$ such that a trajectory starting from $(\tilde{u}_{0,\beta}, 0)$ involves only sliding phases to the left and reaches the equilibrium in finite time, exists if $T \leq T_\beta/2$. Third, any trajectory starting from $(u_0, 0)$ with $u_0 \in]U_d, \tilde{u}_{0,\alpha}[$ either enters the interval $[\tilde{u}_{0,\alpha}, \tilde{u}_{0,\beta}]$ in finite time if $T \leq T_\beta/2$, or enters the interval $[\tilde{u}_{0,\alpha}, U_e[$ in finite time if not. From Step 2, this implies the generic convergence at infinity to U_e. Figure 5.26 represents an example of such a convergence. □

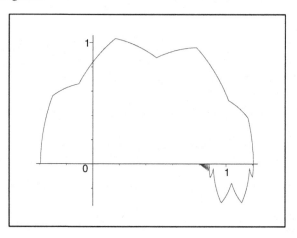

Figure 5.26. *The trajectory enters the interval* $[\tilde{u}_{0,\alpha}, \tilde{u}_{0,\beta}]$ *in finite time and converges toward* U_e *at infinity*

Step 4: The range $T \in]T_\alpha/2, T_\alpha[$ must be divided into the two following parts:

1) $(T_\alpha + T_\beta)/2 \leq T < T_\alpha$: *the trajectory always oscillates around* U_e *and converges to* U_e *at infinity; moreover, the closer* T *is to* T_α, *the slower the convergence is;*

2) $T_\alpha/2 < T < (T_\alpha + T_\beta)/2$: *the trajectory is less smooth, but still converges to* U_e.

The proof will be given only in the first case.

PROOF.– The proof is based on the fact that for T larger than $(T_\alpha + T_\beta)/2$, the motion stays at rest during the time interval $[(T_\alpha + T_\beta), 2T]$ up to the time when the perturbation is set to ε_0 again. This implies that each loop starts from an equilibrium position with a zero velocity, and this situation does not depend on the amplitude so that no phase difference accumulates from one loop to the next and only the value $T_\alpha - T$ determines the rate of convergence toward U_e. The trajectory is represented in Figure 5.27(a). This can be written explicitly in the following way: assume the initial data are $(u_0, 0)$ and the trajectory starts a sliding phase to the right at time $t = 0$ when the perturbation is set to ε_0. Then, the trajectory is built piece wisely as the solution to the following system:

$$
\begin{cases}
\ddot{v}_1 + \omega_\alpha^2 v_1 = F_t + \varepsilon_0 - \mu F_n, \quad t \in]0, T[, \\
v_1(0) = u_0, \quad \dot{v}_1(0) = 0, \\
\\
\ddot{v}_2 + \omega_\alpha^2 v_2 = F_t - \mu F_n, \quad t \in]T, \hat{t}[, \\
v_2(T) = v_1(T), \quad \dot{v}_2(T) = \dot{v}_1(T), \\
\text{with } \hat{t} \text{ such that } \dot{v}_2(\hat{t}) = 0, \\
\\
\ddot{v}_3 + \omega_\beta^2 v_3 = F_t + \mu F_n, \quad t \in]\hat{t}, \hat{t} + T_\beta[, \\
v_3(\hat{t}) = v_2(\hat{t}), \quad \dot{v}_3(\hat{t}) = 0,
\end{cases}
\qquad [5.37]
$$

where the part v_3 involves only sliding to the left while the external force remains equal to zero. The end of this phase is a rest position up to the time $2T$ when the perturbation is set to ε_0 again. Through classical elementary calculations $v_2(\hat{t})$ is expressed as a function of u_0, and the arrival position of the loop is:

$$
v_3(\hat{t} + T_\beta) := u_1 = U_e - (v_2(\hat{t}) - U_e). \qquad [5.38]
$$

Of course, these calculations can be changed into those of a loop starting at $(u_k, 0)$ and arriving at $(u_{k+1}, 0)$. For all k, setting $d_k := U_e - u_k$ equation [5.38] gives

$$
d_{k+1} = \left[d_k^2 + \frac{\varepsilon_0^2}{\omega_\alpha^4} - \frac{2\varepsilon_0}{\omega_\alpha^2} d_k \cos \omega_\alpha T \right]^{1/2} - \frac{\varepsilon_0}{\omega_\alpha^2}. \qquad [5.39]
$$

Since $(T_\alpha + T_\beta)/2 \le T < T_\alpha$, $\dfrac{\pi}{2} \le \omega_\alpha T < \pi$, so that equation [5.39] implies that $d_{k+1} < d_k$. The sequence $\{d_k\}$ has a lower bound which is zero, and passing to the limit in equation [5.39], the limit is shown to be equal to zero so that the trajectory converges to U_e at infinity. □

In the second case, that is when $T_\alpha/2 < T < (T_\alpha + T_\beta)/2$, $\dot{u}(2T)$ is not equal to zero. So, instead of having loops that all start at some data $(u_k, 0)$ as in the first case, that are therefore solution of the same system with formally the same initial data, there is now a phase difference that accumulates at each loop between the trajectory and the loading. Nevertheless, slightly more complicated calculations than the previous ones prove that the trajectory converges to U_e with a rate of convergence that depends on whether T is close to $T_\alpha/2$ or to $(T_\alpha + T_\beta)/2$. An example of this last case is represented in Figure 5.27(b).

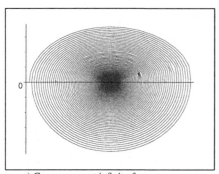
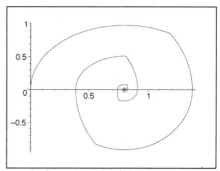

a) Convergence at infinity for	b) Convergence at infinity for
$(T_\alpha + T_\beta)/2 < T < T_\alpha$	$T_\alpha/2 < T < (T_\alpha + T_\beta)/2$

Figure 5.27. *Two trajectories represented in the phase space for $\frac{T_\alpha}{2} < T < T_\alpha$*

5.4. When infinitely many equilibria exist

As stated in section 6.1, the set $\overline{R}_n = [R_{n\varepsilon}^-, R_n^+]$ is a non-zero measure interval when $\varepsilon < \frac{2\mu A}{K_t + \mu W}$. The size of this interval becomes smaller as ε approaches $\frac{2\mu A}{K_t + \mu W}$. Each point R_n^* in this interval corresponds to an equilibrium solution whose tangential component is given by $u_t^* = \frac{R_n^* + F_n}{W}$, which leads to the following result.

PROPOSITION 5.7.– When the amplitude ε of the perturbing force is strictly smaller than $\dfrac{2\mu A}{K_t + \mu W}$:

i) infinitely many equilibrium solutions exist;

ii) all sliding trajectories attain an equilibrium. This equilibrium is in general attained in finite time.

PROOF.– The first point was given in theorem 5.1, indeed any position in the interval $\left[\frac{R_{n\varepsilon}^- + F_n}{W}, \frac{R_n^+ + F_n}{W}\right]$ with a zero velocity satisfies problem [5.1].

The proof of the second point is quite different depending on the value of ε, T and the initial condition u_0. The proof will be detailed in a few of these cases and the different possible trajectories illustrating each case will be shown.

First case: When the initial data are sufficiently close to equilibrium and T is sufficiently large, an equilibrium is reached after only one sliding oscillation to the right and eventually another to the left. This situation is described in the following lemma.

LEMMA 5.4.– If $T \geq T_\alpha$ and

$$Max(\frac{2R_{n\varepsilon}^- - R_{n\varepsilon}^+ + F_n}{W}, \frac{3R_{n\varepsilon}^- - 2R_n^+ + F_n}{W}) \leq u_0 < \frac{R_{n\varepsilon}^- + F_n}{W},$$

all the trajectories issued from $(u_0, 0)$ will attain an equilibrium solution before the end of the period.

PROOF.– The solution of the following system describes a first phase of sliding to the right until the velocity is zero, a second phase during which the particle stays at rest until the end of the half-period, a third phase of sliding to the left and a last phase of rest until the end of the period.

$$\begin{cases} T > T_\alpha, \quad \dfrac{2R_{n\varepsilon}^- - R_{n\varepsilon}^+ + F_n}{W} \leq u_0 < \dfrac{R_{n\varepsilon}^- + F_n}{W}, \\[2mm] \ddot{u}_1 + \omega_\alpha^2 u_1 = F_t - \mu F_n + \varepsilon, \quad t \in (0, T_\alpha) \\ u_1(0) = u_0, \quad \dot{u}_1(0) = 0, \\[2mm] u_2(t) = u_1(T), \quad \dot{u}_2(t) = 0, \quad t \in (T_\alpha, T) \\[2mm] \ddot{u}_3 + \omega_\beta^2 u_3 = F_t + \mu F_n, \quad t \in (T, T + T_\beta) \\ u_1(T) = u_3(T), \quad \dot{u}_3(T) = 0, \\[2mm] \dot{u}_4(t) = 0, \quad t \in (T + T_\beta, 2T). \end{cases} \qquad [5.40]$$

If $u_0 \in [\dfrac{2R_{n\varepsilon}^- - R_n^+ + F_n}{W}, \dfrac{R_{n\varepsilon}^- + F_n}{W}[$, then $u_1(T) \in \left[\dfrac{R_{n\varepsilon}^- + F_n}{W}, \dfrac{R_n^+ + F_n}{W}\right]$ so that $u_2 \equiv u_3 \equiv u_4$ and the result is trivial. The trajectory is shown in Figure 5.28(a).

If $u_0 < \dfrac{2R_{n\varepsilon}^- - R_n^+ + F_n}{W}$, then inserting the solution of [5.40] into the expression of the normal component of the reaction given by the equilibrium equations gives:

$$R_n(2T) - R_n(0) = 2\frac{2\mu WA - \varepsilon W(K_t + \mu W)}{K_t^2 - \mu^2 W^2} = 2(R_n^+ - R_{n\varepsilon}^-),$$

which means that $R_n(2T) \in \overline{R}_n$. Figure 5.28(b) represents such a trajectory. $\square$

REMARK 5.7.– The set of equilibrium solutions are represented in the figures of this section by a thick line on the u axis.

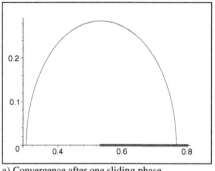
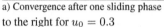
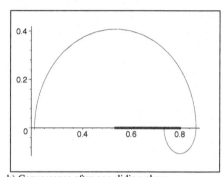

a) Convergence after one sliding phase   b) Convergence after one sliding phase
to the right for $u_0 = 0.3$ to the right and another to the left for $u_0 = 0.2$

Figure 5.28. *Two trajectories represented in the phase space for*
$\varepsilon = 0.8$ *and* $T = 3 \geq T_\alpha$

Second case: Let us now look at the case where u_0 does not belong to the interval given above but where the half-period T is still greater than T_α. In this case, the trajectory does not enter the interval of equilibrium states after a sliding motion to the right followed by a sliding motion to the left but if u_1 represents the tangential position with zero velocity corresponding to the end of such a motion then the following inequality holds:

$$u_1 > u_0 + 2(u^+ - u^{\varepsilon-}),$$

where $u^+ = \frac{R_n^+ + F_n}{W}$ is the upper bound of the interval of equilibrium states and $u^{\varepsilon-} = \frac{R_{n_\varepsilon}^- + F_n}{W}$ is the lower bound of the same interval. So, the following result can be stated:

LEMMA 5.5.– If $T \geq T_\alpha$, a trajectory issued from an initial position $(u_0, 0)$ will attain an equilibrium state after at most k periods where k is the integer part of
$$\frac{u^{\varepsilon-} - u_0}{2(u^+ - u^{\varepsilon-})}.$$

Figure 5.29 represents a trajectory obtained with a value of the amplitude ε close to the critical value ε_0 so the set of equilibrium solutions is relatively small and therefore the trajectory must go through a number of loops before reaching to an equilibrium state.

In the first and second cases, where $T \geq T_\alpha$, all the trajectories attain an equilibrium solution in finite time.

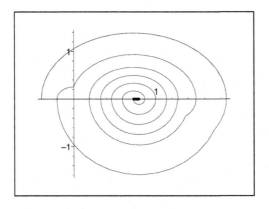

Figure 5.29. *A trajectory attaining an equilibrium*
after six periods for $T \geq T_\alpha$

Third case: What happens when $T < T_\alpha$? In this case, an equilibrium state can be reached in finite time after a phase of successive sliding motions always in the same direction if $T < T_\alpha/3$ or after sliding phases to the right and to the left if $T_\alpha/3 < T < T_\alpha$. The trajectories are very similar to those exhibited in section 6.3.2. However, as was the case in section 6.3.2, there is a value $\bar{u}$ such that if the initial condition is greater than $\bar{u}$ then there are only sliding phases that pile up to the right of $u^{\varepsilon-}$. The following lemma can then be given.

LEMMA 5.6.– If the initial data $(u_0, 0)$ are such that u_0 is strictly smaller than $u^{\varepsilon-}$ and greater than $\bar{u}$ defined by:

$$\bar{u} = u^{\varepsilon-} - \frac{\varepsilon_0 \cos \omega_\alpha T(1 + \tan^2 \omega_\alpha T)}{2\omega_\alpha^2}, \quad \text{when } T \in \,]0, T_\alpha/3[\qquad [5.41]$$

$$\bar{u} = u^{\varepsilon-} - \frac{2\varepsilon_0 \cos \omega_\alpha T}{\omega_\alpha^2}, \quad \text{when } T \in [T_\alpha/3, T_\alpha/2] \qquad [5.42]$$

then the trajectory involves only sliding phases to the right, which pile up at the left of $u^{\varepsilon-}$.

The proof of this lemma is identical to the proof of lemma 5.3 in which U_e has been replaced by $u^{\varepsilon-}$ and therefore will be omitted. An example of such a convergence is given in Figure 5.30.

If the initial condition is taken smaller to the one given by lemma 5.6, then trajectories such as those represented in Figure 5.31 can be obtained. □

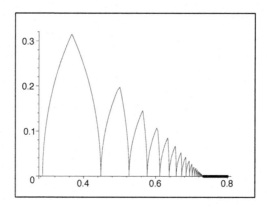

Figure 5.30. *A trajectory attaining an equilibrium at infinity for* $\varepsilon = 1.1$ *and* $T = 0.5 \leq T_\alpha/2$ *where the initial condition* $\bar{u}$ *is given by lemma 5.6*

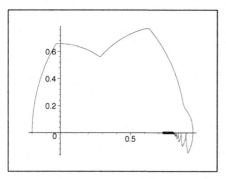
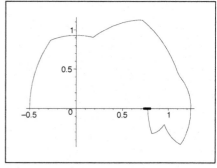

a) Convergence at infinity for $u_0 = -0.2$ b) Convergence in finite time for $u_0 = -0.5$

Figure 5.31. *Two trajectories attaining an equilibrium represented in the phase space for* $\varepsilon = 1.1$ *and* $T = 0.5 \leq T_\alpha/2$

5.5. Toward a more general excitation

This short concluding section can be considered, on the one hand, as an introduction to the generic character of the results of this chapter, and on the other hand as a useful complement to the qualitative analysis presented. The loadings $F_t(t)$ and $F_n(t)$ are still taken of the form:

$$F_t(t) = F_t + P_t(t) \text{ and } F_n(t) = F_n + P_n(t),$$

where $P_t(t)$ and $P_n(t)$ are, respectively, a tangential perturbation and a normal one, and as before only a tangential perturbation is applied, but now this tangential perturbation is equal to $P_t(t) = \varepsilon \sin \gamma t$. Of course, an explicit analytical investigation, as was performed in previous sections, would be quite out of reach for any general loading and in particular for this harmonic one. This section therefore essentially contains results of simulations obtained by symbolic or numerical calculations. The existence of equilibrium states is established in the same way as in the case of the rectangular wave shape excitation.

5.5.1. Existence of equilibrium states

In the case of a sinusoidal perturbation, the set $\{R_n\}(t)$ of normal components of the reaction corresponding to equilibrium solutions is given by:

$$\{R_n\}(t) = \left[\frac{-A + \varepsilon W \sin \gamma t}{K_t - \mu W}, \frac{-A + \varepsilon W \sin \gamma t}{K_t + \mu W} \right].$$

The set $\overline{R}_n$ representing the set of normal components of the reaction at equilibrium for all time is defined by:

$$\overline{R}_n = \bigcap_{t>0} \{R_n\}(t),$$

and in this case, $\overline{R}_n$ becomes:

$$\overline{R}_n = \left[\frac{-A + \varepsilon W}{K_t - \mu W}, \frac{-A - \varepsilon W}{K_t + \mu W} \right].$$

[5.43]

So, $\overline{R}_n$ is an empty set if $\varepsilon > (\mu A)/K_t$, is reduced to a single point when $\varepsilon = (\mu A)/K_t$ and is a non-zero measure interval when $\varepsilon < (\mu A)/K_t$. In other words, through lemma 5.1 there are no equilibrium solutions when $\varepsilon > (\mu A)/K_t$, there is a unique equilibrium solution when $\varepsilon = (\mu A)/K_t$ and there are infinitely many equilibrium solutions when $\varepsilon < (\mu A)/K_t$. The part of the $\{period, amplitude\}$ plane where the trajectories do not lose contact is divided into two parts, the lower one where no periodic solutions exist but infinitely many equilibrium states do, and the upper one, where periodic solutions exist but equilibrium states no longer do.

For example, in the case of the numerical values given in equation [5.5] the transition is obtained for $(\mu A)/K_t = 0.75$, which would have been the value obtained with a rectangular wave force oscillating between $-\varepsilon$ and $+\varepsilon$.

5.5.2. *When infinitely many equilibrium solutions exist ($\varepsilon < (\mu A)/K_t$)*

In this case, the set $\overline{R}_n$ is a non-zero measure interval. The existence of an infinity of equilibrium solutions, as stated above, is a straightforward consequence of lemma 5.1 no matter the perturbation, whether of rectangular or sinusoidal wave shape. But it is interesting, both from the point of view of a constructive proof of the existence of equilibrium states and from that of specific notions of stability, to look at the trajectories starting from any initial data out of equilibrium. It is observed that all trajectories converge toward an equilibrium state in general in finite time. The proof of this result is difficult to establish because in the case of a sinusoidal excitation closed form calculations of the trajectories are no longer possible. However, the following proposition establishes the convergence of the trajectory toward an equilibrium state in the specific case where the convergence is obtained after a single oscillation.

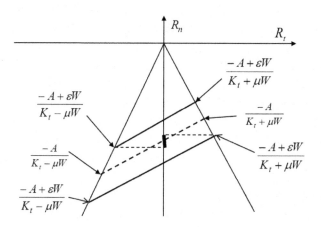

Figure 5.32. *Equilibria for the sinusoidal perturbation: reactions corresponding to non-perturbed equilibrium solutions are on the dashed line – reactions corresponding to perturbed equilibrium solutions are between the two dotted lines*

PROPOSITION 5.8.– If ε is sufficiently small and the frequency γ belongs to $[0.18\omega_\alpha, 1.76\omega_\alpha]$, then the trajectory issued from an initial condition in imminent sliding to the right ($u_0 = \dfrac{F_t - \mu F_n}{K_t - \mu W}$, $v_0 = 0$) attains an equilibrium state after one sliding oscillation to the right.

PROOF.– The amplitude of the perturbation, ε, is chosen small enough (smaller than $\dfrac{\mu A}{K_t}$) to ensure that the interval $\left] \dfrac{-A + \varepsilon W}{K_t - \mu W}, \dfrac{-A - \varepsilon W}{K_t + \mu W} \right[$ is not empty. This interval is represented by a thick line on the R_n axis in Figure 5.32. It is first assumed that $\gamma \neq \omega_\alpha$. The solution after applying the perturbation P_t is:

$$u_t(t) = \frac{F_t - \mu F_n}{K_t - \mu W} + \frac{\varepsilon}{\omega_\alpha(\omega_\alpha^2 - \gamma^2)}(\omega_\alpha \sin(\gamma t) - \gamma \sin(\omega_\alpha t)),$$

until the mass stops sliding at $t^* = \dfrac{2\pi}{\omega_\alpha + \gamma}$ and at that point the normal reaction will be equal to:

$$R_n^* = \frac{-A}{K_t - \mu W} + \frac{\varepsilon W}{K_t - \mu W} \frac{\omega_\alpha}{\omega_\alpha - \gamma} \sin(\frac{\gamma 2\pi}{\omega_\alpha + \gamma}).$$

So, the motion will cease if

$$\frac{-A + \varepsilon W}{K_t - \mu W} < R_n^* < \frac{-A - \varepsilon W}{K_t + \mu W}.$$

A simple computation shows that if γ belongs to $[0.18\omega_\alpha, 1.76\omega_\alpha]$, then $\dfrac{-A + \varepsilon W}{K_t - \mu W} < R_n^*.$

In the case when $\gamma = \omega_\alpha$, the solution is given by:

$$u_t(t) = \frac{F_t - \mu F_n}{K_t - \mu W} - \frac{\varepsilon \omega_\alpha t \cos(\omega_\alpha t)}{2(K_t - \mu W)} + \frac{\varepsilon \sin(\omega_\alpha t)}{2(K_t - \mu W)},$$

and this time when the mass stops sliding its normal reaction is equal to:

$$R_n^* = \frac{-A + \varepsilon W \pi/2}{K_t - \mu W}.$$

So, all values of the frequency of the perturbing force in $[0.18\omega_\alpha, 1.76\omega_\alpha]$ (including $\gamma = \omega_\alpha$) lead to a stop at a time t^* where the reaction is such that:

$$R_n^* > \frac{-A + \varepsilon W}{K_t - \mu W}. \tag{5.44}$$

Moreover, if the amplitude of the perturbation ε is such that:

$$\varepsilon < \frac{2A\mu}{K_t - \mu W + (K_t + \mu W)\dfrac{\omega_\alpha}{\omega_\alpha - \gamma} \sin(\dfrac{\gamma 2\pi}{\omega_\alpha + \gamma})}, \quad \text{when } \omega_\alpha \neq \gamma,$$

$$\text{and } \varepsilon < \frac{2A\mu}{K_t - \mu W + (K_t + \mu W)\dfrac{\pi}{2}}, \quad \text{when } \omega_\alpha = \gamma,$$

then:

$$R_n^* < \frac{-A - \varepsilon W}{K_t + \mu W}.$$

Therefore, when a periodic perturbation is applied for certain values of the frequency of the perturbation, the mass stops after just one phase of sliding to the right and then the mass will stay motionless for all future time if ε is sufficiently small. For all the other values of the frequency of the perturbation, the mass either has a certain number of sliding phases to the right before stopping (as in the case of small periods for the rectangular wave shape perturbations) or slides alternatively to the right and to the left before stopping if ε is sufficiently small. This can be checked, for instance, by computing the solution through a Maple software. □

Numerical simulations have been performed for different values of the frequency γ. The qualitative behavior represented in Figure 5.33 is very close to that of the case where the perturbing load was a rectangular wave, studied extensively in this chapter.

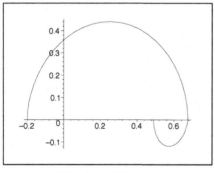 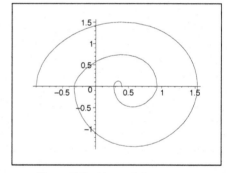

a) $\varepsilon = 0.6$ and $\gamma = 0.4$ b) $\varepsilon = 0.6$ and $\gamma = 1.2$

Figure 5.33. *Evolution in the phase space* $(u_t, \dot{u}_t)$ *when* $\varepsilon = 0.6$

5.5.3. When no equilibrium solutions exist ($\varepsilon > (\mu A)/K_t$)

In this case, there no longer exist any stationary solutions. However, different types of periodic solutions are observed.

Again, this situation is close to the case where the external perturbation was a rectangular wave, as shown in Figures 5.34(a) and (b). Nevertheless, some differences appear since the trajectory represented in Figure 5.35(a) is slightly more complicated than what had been obtained in the previous sections for the same values in the $\{period, amplitude\}$ plane. Moreover, the complexity seems to increase for ε sufficiently large and for large periods, which had never been observed with the rectangular wave. An example is given in Figure 5.35(b).

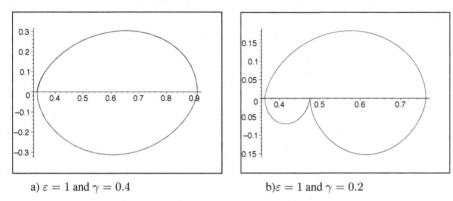

a) $\varepsilon = 1$ and $\gamma = 0.4$ b)$\varepsilon = 1$ and $\gamma = 0.2$

Figure 5.34. *Evolution in the phase space $(u_t, \dot{u}_t)$ when $\varepsilon = 1$*

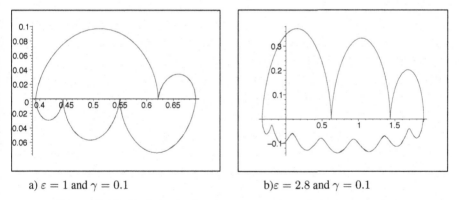

a) $\varepsilon = 1$ and $\gamma = 0.1$ b)$\varepsilon = 2.8$ and $\gamma = 0.1$

Figure 5.35. *Evolution in the phase space $(u_t, \dot{u}_t)$ for large periods*

5.5.4. The transition where a single equilibrium state exists $(\varepsilon = (\mu A)/K_t)$

Here, it seems that there is a qualitative difference with the case of the rectangular waves. The main result of this numerical investigation is the following.

PROPOSITION 5.9.– Whatever the frequency of the oscillating perturbation, the trajectory tends to the single equilibrium at infinity.

Again, Figures 5.36(a) and (b) look very much like those obtained in section 6.3. But here, whatever the value of T on the transition line, only these two different types of trajectories are observed. In the case of the rectangular wave, these trajectories represent the qualitative dynamics only when $T < T_\alpha$. When $T > T_\alpha$, a rectangular wave leads to trajectories that no longer converge to the single equilibrium. They are either periodic or converge toward a periodic trajectory. This result of non-existence of periodic solutions on the transition line in the case of a sinusoidal excitation must be handled with care since it only follows from numerical experiments for the moment, but it is an important qualitative point.

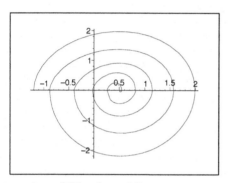
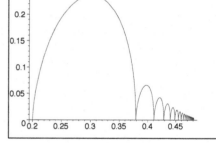

a) $\varepsilon = 0.75$ and $\gamma = 1.2$ b) and $\gamma = 2.4$

Figure 5.36. *Evolution in the phase space* $(u_t, \dot{u}_t)$ *when* $\varepsilon = 0.75$

The Case of the Nonlinear Restoring Force

6.1. Introduction

This chapter investigates the dynamics of the simple mass-spring system when the restoring force is nonlinear but still involves non-regularized unilateral contact and Coulomb friction. As in the previous chapter, the response of the system when submitted to an oscillating excitation will be studied. The main qualitative differences with the case of a linear restoring force are due to the shape of the set of equilibrium states. As classically done in qualitative analysis of dynamical systems, this chapter aims at an investigation of the $\{period, amplitude\}$ plane of the excitation.

The dynamical system studied throughout this chapter was given in Chapter 2 and is recalled here:

$$
\left\{
\begin{array}{l}
i) \left\{
\begin{array}{l}
m\ddot{u}_t = \mathcal{N}_t(u_t, u_n) + F_t + R_t, \\
m\ddot{u}_n = \mathcal{N}_n(u_t, u_n) + F_n + R_n,
\end{array}
\right. \quad t > 0 \\[4mm]
ii) \ u_t(0) = u_t^0, \ u_n(0) = u_n^0, \ \dot{u}_t(0) = \dot{u}_n(0) = 0 \\[3mm]
iii) \ u_n \leq 0, \ R_n \leq 0, \ u_n R_n = 0, \\[3mm]
iv) \left\{
\begin{array}{l}
R_n = 0 \implies \dot{u}_t \in \mathbb{R}, \\[3mm]
\mu R_n \leq R_t \leq -\mu R_n, \\[3mm]
\text{with} \ \left\{
\begin{array}{l}
|R_t| < -\mu R_n \implies \dot{u}_t = 0, \\
|R_t| = -\mu R_n \implies \exists \lambda \geq 0 \ \text{s.t.} \ \dot{u}_t = -\lambda R_t,
\end{array}
\right.
\end{array}
\right. \\[3mm]
v) \ \text{Let } \tau \text{ such that } u_n(\tau) = 0, \text{ then } \dot{u}_n^+(\tau) = -e\dot{u}_n^-(\tau), \ e \in [0,1].
\end{array}
\right.
\qquad [6.1]
$$

The components $\mathcal{N}_t(u_t, u_n)$ and $\mathcal{N}_n(u_t, u_n)$ of the restoring force calculated in Chapter 2 are given by:

$$
\left\{
\begin{array}{l}
\mathcal{N}_t(u_t, u_n) = -k(X_0 + u_t)\left[1 - \dfrac{\sqrt{X_0^2 + h^2}}{\sqrt{(X_0 + u_t)^2 + (h + u_n)^2}}\right], \\[4mm]
\mathcal{N}_n(u_t, u_n) = -k(h + u_n)\left[1 - \dfrac{\sqrt{X_0^2 + h^2}}{\sqrt{(X_0 + u_t)^2 + (h + u_n)^2}}\right].
\end{array}
\right.
\qquad [6.2]
$$

REMARK 6.1.– At small strains, due to the fact that $\mathcal{N}_t(0,0) = \mathcal{N}_n(0,0) = 0$ and to straightforward calculations these nonlinear terms are found to be, respectively, equal to $-k(\sin^2 \varphi\, u_t + \sin \varphi \cos \varphi\, u_n)$ and $-k(\sin \varphi \cos \varphi\, u_t + \cos^2 \varphi\, u_n)$ where φ represents the angle between the spring and the normal to the obstacle (see [BAS 03]). So, the following familiar equations are obtained:

$$
\left\{
\begin{array}{l}
m\ddot{u}_t + K_t u_t + W u_n = F_t + R_t, \\[2mm]
m\ddot{u}_n + W u_t + K_n u_n = F_n + R_n.
\end{array}
\right.
\qquad [6.3]
$$

6.1.1. The equilibrium solutions considered

The set of equilibrium solutions of problem [6.1] has been completely investigated in Chapter 4. It consists of states in contact and out of contact, which may either coexist or exist separately according to the values of the forces. Since equilibrium states out of contact mean that the model is reduced to a classical nonlinear dynamical system, this study focuses on equilibrium states in contact, which obviously implies some bounds on the external force (F_t, F_n). More precisely, inserting $u_n \equiv 0$ into equations [6.1(i) and (ii)] and looking for equilibria, it was shown in Chapter 4 that u_t and R_n are linked by the following expression:

$$
u_t(R_n) = -X_0 \pm h\sqrt{\dfrac{k^2 h^2 \beta^2}{(F_n + R_n - kh)^2} - 1}, \qquad [6.4]
$$

from which the following relation between the tangential and normal component of the reaction R_t and R_n of the equilibrium solution is obtained:

$$
(R_t + F_t)^2 = (F_n + R_n)^2 \left(\dfrac{k^2 h^2 \beta^2}{(F_n + R_n - kh)^2} - 1\right), \qquad [6.5]
$$

where $\beta := \sqrt{1 + (\frac{X_0}{h})^2}$ is the geometrical constant introduced in Chapter 4.

Finding the equilibrium solutions in contact then amounts to finding the intersection, which obviously depends on the choice of the parameters, of the graph of equation [6.5] with the Coulomb cone. The generic situation for the set of equilibria in contact has been presented in Chapter 4. The coordinates of some representative points introduced for the investigation of the equilibrium states are recalled here:

$$\begin{cases} A\begin{vmatrix} -F_t \\ kh(1 - \beta - \frac{F_n}{kh}), \end{vmatrix} & B\begin{vmatrix} -F_t \\ -F_n, \end{vmatrix} \\[2em] C\begin{vmatrix} -F_t - F_n\sqrt{\dfrac{\beta^2}{(\frac{F_n}{kh} - 1)^2} - 1} \\ 0, \end{vmatrix} \\[2em] D\begin{vmatrix} 0 \\ kh(1 - \frac{F_n}{kh}), \end{vmatrix} & E\begin{vmatrix} -kh(\frac{F_t}{kh} + (\beta^{2/3} - 1)^{3/2}) \\ -kh(\frac{F_n}{kh} - 1 + \beta^{2/3}). \end{vmatrix} \end{cases} \qquad [6.6]$$

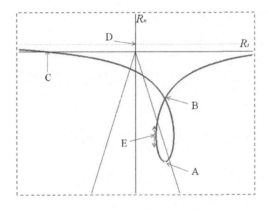

Figure 6.1. *A generic case of the set of equilibria in contact*

The thick curve is the graph given by equation [6.5], that is the equilibrium equation obtained from problem [6.1] in the case $u_n \equiv 0$, and the thin lines delimit the friction cone given by [6.1(iv)]. The analysis presented in Chapter 4 showed that increasing F_t amounts to translating the thick curve to the left, while increasing F_n amounts to pushing the curve downward on the R_n axis. It will also be useful to recall that for any force (F_t, F_n) equation [6.5] defines implicitly the graph represented in Figure 6.1 and that this graph can be divided into two parts having the same horizontal asymptote.

6.1.2. *Recalling a few useful results*

The following result was given in Chapter 3 and was proved in [CHA 14].

LEMMA 6.1.– Under the hypothesis that the restoring force $\mathcal{N}(u)$ is an analytic function of u satisfying a global Lipschitz condition, and that the driving force $F(t)$ is a piecewise analytic function of t, problem [6.1] has a unique solution for any given initial condition that is compatible with the unilateral constraint.

The following technical result, already presented, will be essential for the computation of a trajectory:

LEMMA 6.2.– Let us consider a solution of problem [6.1] under constant loading during a time interval in which contact with the obstacle is always achieved. When the tangential velocity becomes equal to zero after a sliding phase, either the velocity stays equal to zero for all time or the velocity changes sign (i.e. the sliding direction changes).

The proof of this result has been given for lemma 2.1 in Chapter 3. Here, it only requires to specify the sliding equations [2.31] and [2.32] using the adequate nonlinearity, that is:

$$m\ddot{u}_t + k(X_0 + u_t - \mu h)\left[1 - \frac{h\beta}{\sqrt{(X_0 + u_t)^2 + h^2}}\right] = F_t - \mu F_n, \qquad [6.7]$$

for positive sliding, and

$$m\ddot{u}_t + k(X_0 + u_t + \mu h)\left[1 - \frac{h\beta}{\sqrt{(X_0 + u_t)^2 + h^2}}\right] = F_t + \mu F_n, \qquad [6.8]$$

for negative sliding, keeping in mind that in both cases the normal component of the reaction is given by:

$$kh\left[1 - \frac{h\beta}{\sqrt{(X_0 + u_t)^2 + h^2}}\right] - F_n = R_n. \qquad [6.9]$$

The remaining part of the proof has been given in Chapter 3:

REMARK 6.2.– At this point, it is important to stress the difference with the linear case when the trajectories are represented in the $\{R_t, R_n\}$ plane. Indeed, because of the very simple relation between R_n and u_t in the linear case when the mass was sliding to the right it was moving upward on the left side of the cone and conversely when it was sliding to the left it was moving downward on the right side of the cone. In the nonlinear case when the mass is sliding to the right, it is still moving on the left side of the cone (respectively, sliding to the left it is still moving on the right side of the cone) but it can be moving upward or downward depending on its position. This is due to the fact that R_n and u_t do not depend linearly on one another in the nonlinear case (see equation [6.9]).

6.2. A particular case: $F_n > 0$ large enough and $|F_t|$ large enough

The aim of this section is to show that although the set of equilibria is given by a complicated graph and involves a lot of different cases, there are some cases where a good approximation of the dynamics can be obtained by the same calculations as those of the linear case. It has been checked in remark 6.1 that under the hypothesis of small displacements, the equations of the dynamics reduce to the classical equations [5.1]. So, at small strains this linearization leads exactly to the cases that have been explored in Chapter 6. However, the shape of the graph of Figure 6.1 suggests that more interesting linearizations can be undertaken.

6.2.1. Linearization around an equilibrium $(u_t^0, 0)$

The normal component of the external force F_n is taken such that $F_n \geq kh$, which ensures that the horizontal asymptote of the graph of the equilibrium solutions given by equation [6.5] is in the negative half-plane. The point $M := \left(R_t = 0, R_n = -\frac{F_n}{2}\right)$ of the $\{R_t, R_n\}$ plane, which is in the Coulomb cone for any value of μ, is then considered and the values (F_t^0, F_n^0) ensuring that point M is on the graph of the equilibrium solutions are computed. The following values are obtained, $F_n^0 = kh$, $F_t^0 = \frac{kh}{2}\sqrt{4\beta^2 - 1}$ and let $(u_t^0, u_n^0 = 0)$ be the associated equilibrium solution of problem [6.1] with the reaction (R_t^0, R_n^0), that is $u_t^0 = -X_0 + h\sqrt{4\beta^2 - 1}$.

Assume that a small change $\tilde{F}_t$ of the tangential component of the external force modifies the tangential motion from u_t^0 to $u_t^0 + \tilde{u}_t$ and the reaction R_t^0, R_n^0 to $R_t^0 + \tilde{R}_t$, $R_n^0 + \tilde{R}_n$. Going through Taylor expansions, identifying the terms in equilibrium with (F_t^0, F_n^0), and removing higher order terms, this linearization of equations [6.1(i) and (ii)] around point M reads:

$$
\begin{cases}
i) \ m\ddot{\tilde{u}}_t + K_t \tilde{u}_t = \tilde{F}_t + \tilde{R}_t, \\[2mm]
ii) \ W\tilde{u}_t = F_n + \tilde{R}_n,
\end{cases}
\qquad [6.10]
$$

where the coefficients K_t and W are given by the partial derivatives with respect to u_t of $\mathcal{N}_t(u_t, u_n)$ and $\mathcal{N}_n(u_t, u_n)$ at $(u_t^0, 0)$, that is:

$$K_t = k\left(-1 + \frac{1}{8\beta^2}\right), \; W = -\frac{k}{8\beta^2}\sqrt{4\beta^2 - 1}.$$

These coefficients are those of a stiffness matrix that is the linear approximation of the nonlinear restoring force $(\mathcal{N}_t(u_t, u_n), \mathcal{N}_n(u_t, u_n))$ in the neighborhood of point $(u_t^0, 0)$. Figure 6.2 represents the equilibrium solutions with a nonlinear restoring force and its linearization in the neighborhood of point $(u_t^0, 0)$.

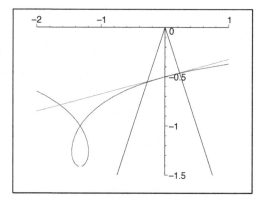

Figure 6.2. *An example of a situation where the set of equilibria in contact is close to a straight line*

REMARK 6.3.– It is important to keep in mind that linearizing the whole problem [6.1] would be meaningless, since the right-hand sides of lines (i) and (ii) involve measures and lines (iii) and (iv) involve inequalities.

Throughout this chapter, "linear" or "linearized" will be used without ambiguity for the problem where the equations of the dynamics have been linearized but where the nonlinearities due to the contact and friction constraints remain. The problem will be referred to as the "nonlinear" problem when both the equations of the dynamics and the contact and friction constraints are nonlinear. In the remaining part of section 7.2, it is shown that solving the linearized problem gives a very good approximation of the solution to the nonlinear one when the situation corresponds to the one represented in Figure 6.2. Unfortunately, this is not always the case, so the section ends by explaining that this linearization must be used with caution, emphasizing the situations where nonlinear and linearized solutions differ.

6.2.2. *Approximating the qualitative dynamics*

In section 6.2.1, we gathered that in order to calculate the sliding trajectories around point M, problem [6.1] can be replaced by a linearized problem, which involves equations [6.10]. The dynamics of this linearized problem has already been completely investigated. Two examples comparing the behavior of the solution of the initial nonlinear system with that of the solution of equations [6.10] with the same initial data and unilateral contact and friction conditions are given. In both cases, a periodic perturbation of amplitude ε and period $2T$ is applied to the tangential force. The trajectories represented in Figure 6.3 go to equilibrium in finite time and it is interesting to observe the proximity of the final times and of the positions at arrival of the nonlinear problem and its linearization. Figure 6.4 contains periodic solutions obtained for a perturbing force of larger amplitude than that of Figure 6.3.

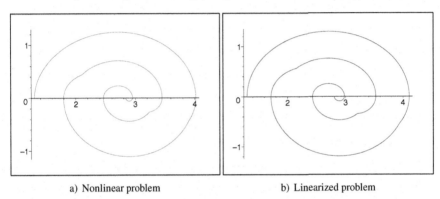

a) Nonlinear problem   b) Linearized problem

Figure 6.3. *Comparison of the convergence to an equilibrium solution for the fully nonlinear problem and its linearization when $T = 4$ and $u_t(0) = 1.3$. The nonlinear equilibrium $u_t = 2.832528624$ is obtained at $t = 23.107$ while the trajectory reaches the equilibrium $u_t = 2.801836470$ at time $t = 23.110$ using the linearized equations*

REMARK 6.4.– The algorithms used to compute the trajectories represented in Figures 6.3 and 6.4 were given in Chapter 6 for the linear case and will be given in the next section for the nonlinear case.

A few comments about this approximation can now be added:

– this analysis will hold for any normal component $F_n \geq hk$; it can be transposed if F_t is changed into $-F_t$ by symmetry with respect to the R_n axis in the $\{R_t, R_n\}$ plane. In particular, the symmetry means that the slope of the straight line that approximates the nonlinear graph changes sign, which simply implies that the rotation around the equilibrium changes sign;

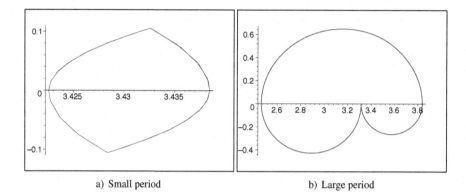

a) Small period   b) Large period

Figure 6.4. *Comparison of periodic trajectories: (a) for a very small period,* $T = 0.3$: *For* $\varepsilon = 1.5$, *a periodic solution is found in the case of the nonlinear strain for an initial value* $(u_0 = 3.4225, \dot{u}_0 = 0)$, *while the linearized problem gives* $(u_0 = 3.2079, \dot{u}_0 = 0)$; *(b) for a large period,* $T = 10$: *For* $\varepsilon = 0.7$, *a periodic solution is found in the case of the nonlinear strain for an initial value* $(u_0 = 2.4767, \dot{u}_0 = 0)$, *while the linearized problem gives* $(u_0 = 2.4860, \dot{u}_0 = 0)$

– as $|F_t|$ increases, the quality of the approximation given by problem [6.10] increases because the part of the graph of equation [6.5] that intersects the Coulomb cone becomes close to a straight line;

– through this approximation, the nonlinear system behaves in the same way as the linear one concerning the existence of a transition between a range where there exist infinitely many equilibrium states and no periodic solutions and another range where there exist periodic solutions but no equilibrium states. The study of this transition is one of the topics of the following sections.

6.2.3. *Where the qualitative dynamics of the linearized problem differs from the nonlinear one*

There are two main qualitative differences in the behavior of the nonlinear problem and its linearization. These differences appear even in a small neighborhood of the point around which the linearization has been performed.

6.2.3.1. *Losing contact*

The first important difference in the behavior of the solutions is that for the linear problem, a sufficiently large perturbation provokes a loss of contact, whereas for a whole range of nonlinear problems (in fact as soon as $F_n \geq kh$), no loss of contact appears, however large the perturbation. Indeed, for the linearized system, the set of equilibria is represented by a straight line of non-zero slope that will obviously intersect the R_t axis. But for the nonlinear system, the graph of the set of equilibria in the $\{R_t, R_n\}$ plane never intersects the R_t axis when the horizontal asymptote of the graph of equation [6.5] is in the negative half-plane (as is the case when $F_n \geq hk$).

6.2.3.2. Period doubling solutions

Referring now to Figure 6.4(b), it was shown that in the range of the $\{T, \varepsilon\}$ plane where such a trajectory exists, the linearized system exhibits infinitely many periodic solutions around the one represented in Figure 6.4(b). Moreover, in the linear case, all the periodic trajectories in a neighborhood of the trajectory of Figure 6.4(b) have a period twice that of the excitation. Numerical experiments in the nonlinear case have never exhibited such a set of periodic solutions. On the contrary, non-periodic trajectories are obtained and they all converge to the periodic solution of Figure 6.4(b).

6.3. Qualitative study of the set of equilibria

6.3.1. Changes of the set of equilibria under an increasing load

In order to understand how the set of equilibrium solutions evolves when the load increases the graph of the equilibrium solutions under three different loads in the $\{R_t, R_n\}$ plane has been plotted in Figure 6.5 together with the corresponding equilibria.

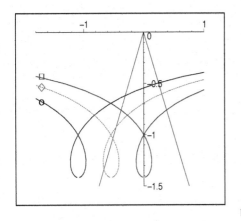
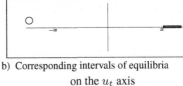

a) Analysis in the $\{R_t, R_n\}$ plane

b) Corresponding intervals of equilibria on the u_t axis

Figure 6.5. *Variation of the set of equilibrium states under increasing tangential load: - box: $F_t = 0$, - diamond: $F_t = F_t^* > 0$, - circle: $F_t = F_t^{**} > F_t^*$*

Figure 6.5(a) shows the modification of the intersection of the graph of equation [6.5] with the friction cone when the graph is translated from right to left, that is under an increasing load. The set of equilibria corresponding to the three different situations is represented in Figure 6.5(b), where the tangential component u_t has been translated in order to have the three trivial equilibrium solutions either side of the u_n axis. In all the remaining figures, this representation has been chosen.

6.3.2. *Some preliminary calculations*

The equilibria around $u_t(R_n) = 0$ are considered (the observations and the analysis would be exactly the same around $u_t(R_n) = -2X_0$), so equation [6.5] gives:

$$R_t(R_n) = -F_t + (R_n + F_n)\sqrt{\frac{h^2 k^2 \beta^2}{R_n^2} - 1}. \qquad [6.11]$$

In order to characterize the different situations that will occur, the following computations are needed. The first step consists of determining, in the case where $F_t = 0$ (that is when the graph of equilibria is centered on the R_n axis) and $F_n = hk$ (that is when the horizontal asymptote of the graph of equilibria is the R_t axis), the value of the friction coefficient that ensures that the graph of equation [6.11] is tangent to the left border of the Coulomb cone. Writing that $R_t(R_n)$ given by equation [6.11] is equal to μR_n, and that their first derivatives are also equal gives:

$$\begin{cases} i)\ \mu R_n = (R_n + kh)\sqrt{\dfrac{h^2 k^2 \beta^2}{R_n^2} - 1} \\[4mm] ii)\ \mu = \dfrac{-R_n^3 - h^3 k^3 \beta^2}{R_n^3 \sqrt{\dfrac{h^2 k^2 \beta^2}{R_n^2} - 1}}, \end{cases} \qquad [6.12]$$

from which, going through straightforward calculations, R_n must be a solution to:

$$R_n^2 - kh\beta^2 R_n - 2h^2 k^2 \beta^2 = 0,$$

whose only negative root is:

$$\widehat{R}_n = \frac{hk\beta^2}{2}\left(1 - \sqrt{1 + \frac{8}{\beta^2}}\right),$$

and the corresponding R_t is given by equation [6.11]. Inserting this value into equation [6.12], the graph of equation [6.11] is found to be tangent to the Coulomb cone if the coefficient of friction is equal to:

$$\mu_c = \left(1 + \frac{kh}{\widehat{R}_n}\right)\sqrt{\frac{h^2 k^2 \beta^2}{\widehat{R}_n^2} - 1}.$$

These values, although intricate, depend only on the given geometry and are easily computed.

From a more practical point of view, under a given geometry and a given friction coefficient the amplitude of the tangential perturbation that leads to the tangency on the left border of the cone is sought for. Again from equation [6.11] now written as:

$$\mu R_n + F_t = (R_n + kh)\sqrt{\frac{h^2 k^2 \beta^2}{R_n^2} - 1},$$

the following expression is obtained:

$$F_t = \frac{kh(-R_n^2 + kh\beta^2 R_n + 2h^2 k^2 \beta^2)}{R_n^2 \sqrt{\dfrac{h^2 k^2 \beta^2}{R_n^2} - 1}}, \qquad [6.13]$$

where the value of R_n at the tangency satisfies equation [6.12(ii)], that is:

$$\mu R_n^3 \sqrt{\frac{h^2 k^2 \beta^2}{R_n^2} - 1} + R_n^3 + h^3 k^3 \beta^2 = 0.$$

From the coordinates of point E given in formula [6.6], the value of R_n at the tangency belongs to the interval $]-hk\beta^{2/3}, 0[$ and as the function [6.11] is monotone in this interval, this equation has a single root in the interval $]-hk\beta^{2/3}, 0[$, so the value of R_n is easily given by a symbolic calculation software. Inserting this value into formula [6.13] gives the amplitude of the tangential component of the force for which the graph of equation [6.11] is tangent to the left border of the cone. The numerical value of this amplitude is denoted by F_{t1}.

The case where the tangency holds on the right border of the cone is now considered. The border of the cone is therefore given by equation $R_t(R_n) = -\mu R_n$ and the equilibrium point at tangency by:

$$-\mu R_n + F_t = (R_n + kh)\sqrt{\frac{h^2 k^2 \beta^2}{R_n^2} - 1}.$$

It is then trivial to see that the same equation as [6.13] is obtained but, as the point of tangency is this time situated between point A and point E due to formula [6.6], the value of R_n at the tangency belongs to the interval $] - hk\beta, -hk\beta^{2/3}[$ and function [6.11] being monotone in this interval, this equation has a single root in the interval $] - hk\beta, -hk\beta^{2/3}[$. This leads to a value of F_t, which is denoted by F_{t2}. Through elementary geometrical properties, it is obvious that $|F_{t1}| < |F_{t2}|$.

These results can be summarized as follows where μ is assumed, without restriction, large enough for the set of equilibria to be simply connected for $F_t = 0$. This gives a quantitative analysis of what was observed in Figure 6.5 keeping in mind that equation [6.11] represents only the right-hand branch of equation [6.5].

$- F_t < F_{t2} < 0$: The graph of equation [6.11] does not intersect the cone, so there is no equilibrium point in contact associated with this branch;

$- F_t = F_{t2}$: The graph of equation [6.11] is tangent to the right border of the cone, so a single equilibrium point in contact appears, which is in imminent sliding to the left;

$- F_{t2} < F_t < F_{t1}$: As F_t increases, the set of equilibrium points in contact increases monotonically;

$- F_t = F_{t1}$: This is where the set of equilibria is the largest while one equilibrium point appears in imminent sliding to the right;

$- F_t > F_{t1} > 0$: A hole appears in the set of equilibria from the point in imminent sliding to the right, and extends monotonically inside the set of equilibria. So, the set of equilibria is no longer connected, which is an important qualitative difference with the case of linear stiffness.

It is trivial that this analysis applies to the equilibria associated to the other branch

$$R_t(R_n) = -F_t - (R_n + F_n)\sqrt{\frac{h^2 k^2 \beta^2}{R_n^2} - 1} \qquad [6.14]$$

of the set of equilibria in contact after a symmetry with respect to the R_n axis.

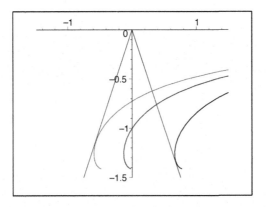

Figure 6.6. *The three limit cases of equilibrium sets. On the left $F_t = F_{t1} > 0$, in the middle $F_t = 0$ and on the right $F_t = F_{t2} < 0$*

6.3.3. *Equilibria under tangential oscillations*

From now on, the normal component $F_n(t)$ of the loading is kept constant, equal to or larger than hk. This insures that there always exist equilibrium solutions in contact whatever the tangential component (see Chapter 4). In fact, $F_n(t) = hk$ is chosen from now on. The tangential component $F_t(t)$ is written in the following way:

$$F_t(t) = F_t + P_t(t),$$

and, as was the case in Chapter 6, a rectangular wave shape perturbation will be applied, defined, for $i = 0$, 1, ... and $\varepsilon > 0$, by:

$$P_t(t) = \varepsilon \text{ if } t \in \ ]2iT, (2i+1)T] \text{ and } P_t(t) = 0 \text{ if } t \in \ ](2i+1)T, (2i+2)T]. \quad [6.15]$$

6.3.3.1. *The set of equilibria*

In the nonlinear case, explicit closed-form calculations cannot be performed as far as in the linear case but the set of equilibrium solutions under such a perturbation is nevertheless very easy to compute. The sets of equilibrium solutions submitted to an oscillation are given by:

$$\overline{\Sigma} = \bigcap_{t>0} \Sigma(t),$$

where $\Sigma(t)$ consists of the set of tangential components of the equilibrium positions at time t, their normal components being equal to zero as these equilibria are in contact. Due to the fact that $F_n(t) = hk$, the set $\Sigma(t)$ is never empty. In the case of a rectangular wave shape perturbation given by [6.15], $\Sigma(t)$ is constant equal to Σ_0 for $t \in](2i + 1)T, (2i + 2)T]$, and constant equal to Σ_ε for $t \in]2iT, (2i + 1)T]$, so $\overline{\Sigma}$ reduces to:

$$\overline{\Sigma} = \Sigma_0 \cap \Sigma_\varepsilon.$$

Therefore, $\overline{\Sigma}$ may consist either of an interval, of several disjoint intervals, of isolated points or of an empty set, depending on the value of ε. For example, referring to Figure 6.5(b), there are two disjoint sets of equilibria corresponding to the intersection of the "box" curve and the "diamond" curve, a single equilibrium point corresponding to the intersection of the "circle" curve and the "diamond" curve or an empty set corresponding to the intersection of the "box" curve and the "circle" curve.

6.3.3.2. Transition to non-existence of equilibria

As long as the set of equilibria $\overline{\Sigma}$ is non-empty, there exist equilibrium solutions for the oscillating load. But the measure of $\overline{\Sigma}$ is maximum for $\varepsilon = 0$ and decreases monotonically to zero as ε increases. Therefore, $\overline{\varepsilon}$ can be defined as the amplitude of the perturbing force for which $\overline{\Sigma}$ reduces to isolated points. For any $\varepsilon > \overline{\varepsilon}$, $\overline{\Sigma} = \emptyset$. In other words, for $\varepsilon > \overline{\varepsilon}$, there no longer exist any equilibrium solutions. The value of $\overline{\varepsilon}$ depends on all the parameters of the problem, its computation is rather intricate and will be omitted.

6.4. Qualitative dynamics when ε is smaller than $\overline{\varepsilon}$

All the figures and numerical results presented in the following sections have been obtained with the following values of the parameters:

$$m = 1, h = 1, X_0 = 1, k = 1 \text{ and } \mu = 0.5.$$

6.4.1. Comments about the calculations

Numerical calculations are presented, either in order to illustrate theoretical results, or simply as numerical experiments when theoretical results are not yet available. The manner in which these calculations were performed is described and justified below. Both the nonsmoothness due to frictional contact and the nonlinearity of the dynamics must be handled with extreme care.

From a practical point of view, by introducing $y := \dfrac{u_t + X_0}{h}$, equations [6.7] and [6.8] become:

$$
\begin{cases}
(i) \;\; \ddot{y} + \dfrac{k}{m}(y - \mu)\left[1 - \dfrac{\beta}{\sqrt{y^2 + 1}}\right] = \dfrac{F_t - \mu F_n}{mh}, \\[3mm]
(ii) \;\; \ddot{y} + \dfrac{k}{m}(y + \mu)\left[1 - \dfrac{\beta}{\sqrt{y^2 + 1}}\right] = \dfrac{F_t + \mu F_n}{mh},
\end{cases}
\qquad [6.16]
$$

so the classical phase plane $\{u_t, \dot{u}_t\}$ is changed into the $\{y, \dot{y}\}$ plane (see from Figure 6.7 onward). It is in fact easier to read the figures in the $\{y, \dot{y}\}$ plane because $y = 0$ corresponds to the case where the spring is vertical.

The existence and uniqueness result given in lemma 6.1 ensures that given initial data, a numerical scheme provides an approximation of the solution to these equations. A fourth-order Runge–Kutta method was used to integrate [6.16] while symbolic calculations were used to check that a sliding phase holds to the left or to the right. In fact, the basic structure of the algorithm used to solve this problem is the same as in the case of linear strains, the only difference is that in the latter case a closed-form expression was obtained for the solution of these sliding phases, whereas here a numerical scheme must be used. Note that very smooth nonlinear differential equations govern both sliding phases so the error estimates on the results presented are close to 10^{-12}.

When the loading is constant, lemma 2.1 determines the trajectory after a time t^* at which the velocity is equal to zero: either the velocity stays equal to zero for all future time, that is the mass has attained an equilibrium and is clamped by friction, or the velocity changes sign, that is the mass starts sliding in the opposite direction. When the loading is oscillating due to a perturbation of the form [6.15], lemma 2.1 also determines the trajectory after a time t^* at which the velocity is equal to zero, but only during a half-period where the loading is constant. If the velocity has remained equal to zero for some time before the end of a half-period, that is the mass has been at equilibrium for that value of the loading, how can the sign of the future velocity be determined when the load changes at the beginning of the next half-period. Does the mass stay motionless or does it start sliding to the right or to the left? The answer is given by computing the tangential and normal components of the new reaction corresponding to the position of the mass at time t^* with the new load. Coulomb's friction law then implies that:

– if $R_t(t^*) < \mu R_n(t^*)$, then the mass starts sliding to the right;

– if $R_t(t^*) > -\mu R_n(t^*)$, then the mass starts sliding to the left;

– and of course if $\mu R_n(t^*) \leq R_t(t^*) \leq -\mu R_n(t^*)$, the mass stays motionless, which means that an equilibrium solution has been obtained for this oscillating load.

6.4.2. Preliminary: the case of constant loading

Let us start by showing a few trajectories. Figures 6.7, and 6.8(a) and (b) represent sets of trajectories issued from initial data in the four quadrants of the phase plane and corresponding, respectively, to $(F_n = kh, F_t = 0)$, $(F_n = kh, F_t = F_t^*)$ and $(F_n = kh, F_t = F_t^{**})$ as defined in Figure 6.5, where the sets of equilibria are composed of two disjoint sets in Figure 6.8(a) (the diamond case of Figure 6.5(b) or of a single isolated set in Figure 6.8(b) (the circle case of Figure 6.5(b)).

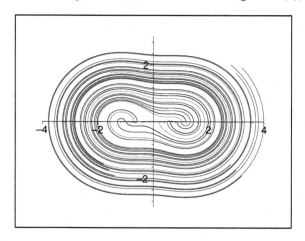

Figure 6.7. *A set of trajectories in the phase space in the case* $F_t = 0$

REMARK 6.5.– It is interesting to note that due to the fact that the set of equilibria is not connected in general, an extreme sensitivity to initial data is observed, as shown in Figure 6.9.

6.4.3. The qualitative dynamics

The study of the dynamics when the tangential loading is oscillating according to formula [6.15] can now be undertaken. The assumption on the normal loading implies that there are always equilibrium solutions in contact if the tangential loading is constant. The first result given here is formally close to the one obtained for the oscillator with a linear stiffness presented in Chapter 6. It can be stated as:

PROPOSITION 6.1.– There exists a value $\bar{\varepsilon}$ of the amplitude of the oscillating load such that, for any $\varepsilon < \bar{\varepsilon}$:

  i) infinitely many equilibrium states exist whatever the period of the excitation;

  ii) all trajectories lead to equilibrium.

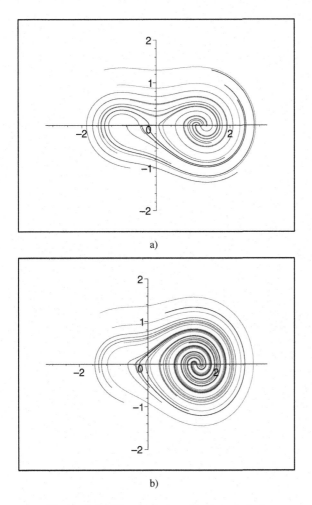

a)

b)

Figure 6.8. *Phase plane under increasing sufficiently small tangential load*

PROOF.– Point (i) is a simple consequence of section 6.3.3. Indeed, $\bar{\varepsilon}$ is defined as the smallest value of ε such that the measure of $\overline{\Sigma} = \Sigma_0 \cap \Sigma_\varepsilon$ is equal to zero. For any value of ε strictly smaller than $\overline{\varepsilon}$, the measure of $\overline{\Sigma}$ is strictly positive so that $\overline{\Sigma}$ consists of one or several segments of equilibria.

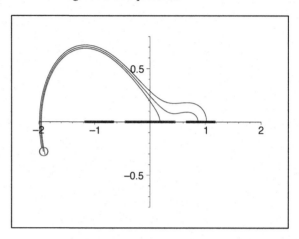

Figure 6.9. *Sensitivity to initial data: trajectories starting from initial data in the same small neighborhood (the small circle in the bottom left quarter of the phase space) diverge and reach very different equilibrium states*

The proof of point (ii) is more intricate. Some basic tools are recalled.

Having a second-order ordinary differential equation in $\mathbb{R}$ of the form:

$$\frac{d^2u}{dt^2} + f(u) = G,$$

where $f(u)$ is some smooth nonlinear function having a primitive $F(u)$, and where G depends neither on t nor on u (although the method could be adapted), a classical integration process, which is just recalled briefly, reads as follows:

Let $\dfrac{d^2u}{dt^2} + f(u) = G,$

then $\dfrac{d^2u}{dt^2}\dfrac{du}{dt} + f(u)\dfrac{du}{dt} = G\dfrac{du}{dt},$

then $\dfrac{1}{2}\dfrac{d}{dt}\left(\dfrac{du}{dt}\right)^2 + \dfrac{d\,F(u)}{dt} = G\dfrac{du}{dt},$

so that $\dfrac{1}{2}\left(\dfrac{du}{dt}\right)^2 + F(u) = Gu + C,$

[6.17]

where C is an integration constant that is given by the initial conditions. Then:

$$\frac{du}{dt} = \pm\sqrt{C + 2(Gu - F(u))},$$

$$\text{and } t = \pm \int_{u_0}^{u} \frac{ds}{\sqrt{C + 2(Gs - F(s))}}.$$

[6.18]

In the present case, the calculations are performed using the functions $f(u)$ given in equations [6.16] successively in the cases of positive sliding and negative sliding, the forcing varying as in formula [6.15]. Recalling the change of variables introduced in equations [6.16] and assuming that sliding holds in the positive direction, the following explicit expression is obtained:

$$\begin{cases} a) \; \dot{y} = \sqrt{C + \frac{2k}{m}\left[\left(\mu + \frac{F_t - \mu F_n}{kh}\right)y - \frac{1}{2}y^2 - \mu\beta \, argsh(y) + \beta\sqrt{1 + y^2}\right]}, \\ \text{and} \\ b) \; t = \int_{y_0}^{y} \frac{ds}{\sqrt{C + \frac{2k}{m}\left[\left(\mu + \frac{F_t - \mu F_n}{kh}\right)y - \frac{1}{2}y^2 - \mu\beta \, argsh(y) + \beta\sqrt{1 + y^2}\right]}}. \end{cases}$$

[6.19]

When the external force is perturbed by a positive tangential component of amplitude ε, according to equation [6.15], only the term $F_t - \mu F_n$ is changed into $F_t + \varepsilon - \mu F_n$. Let the pair (y_0, v_0) be the initial data at $t = 0$, then equation [6.19(a)] gives:

$$\begin{aligned} C = \; & v_0^2 - 2\left(\frac{\mu k}{m} + \frac{F_t - \mu F_n}{mh}\right)y_0 + \frac{k}{m}y_0^2 + 2\frac{\mu k \beta}{m} argsh(y_0) \\ & - 2\frac{k\beta}{m}\sqrt{1 + y_0^2}. \end{aligned}$$

[6.20]

Similar expressions are obtained for negative sliding when the tangential force is equal to $F_t + \varepsilon$ or to F_t.

Then point (ii) of proposition 6.1 will result from the two following lemmas:

LEMMA 6.3.– Let $y_1(t)$ and $y_2(t)$ be two trajectories starting from the same initial data $\{y(0) = y_0, \dot{y}(0) = v_0\}$, $y_1(t)$ being associated with a tangential component of the force $F_t + \varepsilon$ for $\varepsilon < \bar{\varepsilon}$, and $y_2(t)$ being associated with a tangential component of the force F_t.

Then $y_1(t) > y_2(t)$ as long as $\dot{y}_1(t)$ and $\dot{y}_2(t)$ do not reach zero:

PROOF.– Let us start with the case of positive sliding, that is $(y(0) = y_0, \dot{y}(0) = v_0 > 0)$. Then, y_1 is a solution of $(\dot{y}(t))^2 := g_1(y)$ and y_2 is a solution of $(\dot{y}(t))^2 := g_2(y)$, where g_1 and g_2 are given by equation [6.19(a)] in which the tangential component of the force is $F_t + \varepsilon$ for g_1 and F_t for g_2. Thus:

$$g_1(y) - g_2(y) = \frac{2}{m}\varepsilon(y - y_0),$$

which directly gives the result since the right-hand side is positive. The initial data (y_0, v_0) were assumed to be in the positive half-plane. In the case of negative sliding, that is when $(y(0) = y_0; \dot{y}(0) = v_0 < 0)$:

$$g_1(y) - g_2(y) = \frac{2}{m}\varepsilon(y - y_0),$$

which is negative since negative sliding implies $y < y_0$.   □

LEMMA 6.4.– Whatever the external force F_t, let $y_1(t)$ be a solution of equation [6.18] corresponding to positive sliding, which reaches the value $\bar{y}_1$ at zero velocity. Assume $\bar{y}_1$ is out of equilibrium and let $y_2(t)$ be solution of equation [6.18] corresponding to negative sliding starting from the initial data $(\bar{y}_1, 0)$.

Then, $|\dot{y}_1(t)| > |\dot{y}_2(t)|$ as long as $\dot{y}_1(t)$ and $\dot{y}_2(t)$ do not change sign:

PROOF.– The two mappings y_1 and y_2 are defined as being solution to the two differential equations $(\dot{y}(t))^2 := g_1(y)$ and $(\dot{y}(t))^2 := g_2(y)$, where g_1 and g_2 are now defined by:

$$\begin{cases} g_1(y) = \dfrac{2k}{m}\left[\left(\mu + \dfrac{F_t - \mu F_n}{kh}\right)(y - \bar{y}_1) - \dfrac{1}{2}(y^2 - \bar{y}_1^2) \right. \\ \qquad\qquad \left. -\mu\beta(argsh(y) - argsh(\bar{y}_1)) + \beta(\sqrt{1 + y^2} - \sqrt{1 + \bar{y}_1^2})\right] \\[2mm] g_2(y) = \dfrac{2k}{m}\left[\left(-\mu + \dfrac{F_t + \mu F_n}{kh}\right)(y - \bar{y}_1) - \dfrac{1}{2}(y^2 - \bar{y}_1^2) \right. \\ \qquad\qquad \left. +\mu\beta(argsh(y) - argsh(\bar{y}_1)) + \beta(\sqrt{1 + y^2} - \sqrt{1 + \bar{y}_1^2})\right]. \end{cases} \quad [6.21]$$

As long as the trajectories associated with equations [6.21] do not intersect the y axis:

$$\begin{aligned} g_1(y) - g_2(y) &= -\frac{4\mu F_n}{mh}(y - \bar{y}_1) + \frac{4k\mu}{m}(y - \bar{y}_1) \\ &\quad - \frac{4\mu\beta}{m}(argsh(y) - argsh(\bar{y}_1)). \end{aligned}$$

The growth of the function $argsh(.)$ is slower than linear, and β is strictly larger than one, which makes it possible to have an estimate of this difference, but for the contact to hold strictly for any F_t, $F_n = hk$ has been chosen, which gives:

$$g_1(y) - g_2(y) = -\frac{4\mu\beta}{m}(argsh(y) - argsh(\bar{y}_1)).$$

Therefore, $g_1(y) - g_2(y)$ is strictly positive since y is always strictly smaller than $\bar{y}_1$ for both trajectories. $\qquad\qquad\qquad\qquad\qquad\qquad\qquad\qquad\qquad\square$

The trajectories y_1 and y_2 defined in these two lemmas are schematically represented in the phase space in Figure 6.10.

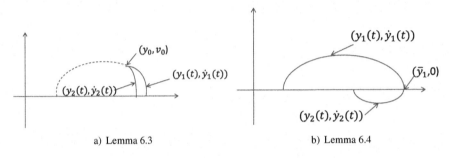

a) Lemma 6.3 $\qquad\qquad\qquad\qquad\qquad\qquad$ b) Lemma 6.4

Figure 6.10. *Qualitative illustration of lemmas 6.3 and 6.4*

These qualitative results are obtained without any integration but only by comparing the right-hand side of two differential equations.

Another intermediate result is now added, which could be seen as a corollary of lemmas 6.3 and 6.4.

LEMMA 6.5.– Assume that under a tangential force $F_t = \varepsilon$, a first phase of positive sliding from an initial data $(y_0, 0)$, denoted by $y_1(t)$, holds up to a point out of equilibrium $(\bar{y}_1, 0)$ at $t = t_1$, where the velocity passes through zero for the first time. The trajectory then continues by a second phase $y_2(t)$ in negative sliding. This trajectory would reach a zero velocity again at some point $\bar{y}_2$ at $t = \bar{t}_2$ if F_t remained unchanged. But at some time t_2 strictly between t_1 and $\bar{t}_2$, the external force F_t is set equal to zero. Let $y_3(t)$ be this last phase, still in negative sliding, and $\bar{y}_3$ its value when the velocity reaches zero again. Then, there exists ε_0 such that:

$$\forall\, t_2 \in\,]t_1, \bar{t}_2[, \quad \begin{cases} \bar{y}_3 > y_0 \ \ if\ \ 0 < \varepsilon < \varepsilon_0, \\[2mm] \bar{y}_3 \leq y_0 \ \ if\ \ \varepsilon \geq \varepsilon_0. \end{cases} \qquad\qquad [6.22]$$

The trajectories described in lemma 6.5 are represented in Figure 6.11 in the case $\varepsilon < \varepsilon_0$. In fact, lemma 6.5 shows that the amplitude of the perturbing force ε controls the arrival point after the complete loop, in particular whether this point is smaller than the initial point y_0 or not. It is therefore the amplitude of the perturbing force ε, which implies that either all the trajectories go towards equilibrium or that the amplitude of the trajectories increases, which in turn allows the existence of periodic solutions.

Since comparing $\dot{y}_1(t)$ to $\dot{y}_2(t)$ and $\dot{y}_2(t)$ to $\dot{y}_3(t)$ has already been undertaken, the calculations are essentially the same as those of lemmas 6.3 and 6.4 but they are rather more tedious so will be omitted.

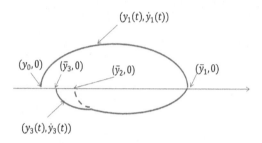

Figure 6.11. *Qualitative illustration of lemma 6.5*

These results can now be used to complete the proof of point (ii) of proposition 6.1:

– let us first assume that the period of the oscillating load is very large. Then, the trajectories behave as if the forcing was constant. During a whole half-period, the calculation involves only lemma 6.4 and the trajectory is represented by any trajectory of Figure 6.7 or 6.8;

– as the period decreases, the calculation of the trajectory cannot avoid taking into account the change of the force during a phase of positive sliding or of negative sliding. Moreover, the smaller the period, the larger the number of such changes. What happens at these changes is qualitatively given by either lemma 6.3 or lemma 6.5;

– calculating a complete trajectory implies accumulating any number of successive phases of positive sliding with the perturbing force or without the perturbing force and of negative sliding with or without the force. This does not seem to lead to new theoretical difficulties, but of course leads to increasing complexity as the period decreases, so the calculations require a symbolic computation software.

Numerical computations suggest that ε_0 defined in lemma 6.5 is equal to the value $\bar{\varepsilon}$ after which no equilibrium solution exists, which was defined at the end of section 7.3.3.2. This implies that the behavior of the system involves a single transition.   □

Trajectories associated with the perturbation of rectangular wave shape defined above are represented in Figure 6.12. Through comparison with Figures 6.7, and 6.8(a) and (b), the non-regular points correspond to the discontinuities of the external perturbation.

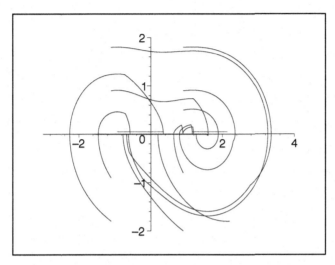

Figure 6.12. *A set of trajectories associated with* $\varepsilon < \overline{\varepsilon}$ *and* $T = 3$

6.5. Periodic solutions when ε is greater than $\overline{\varepsilon}$

The amplitude of the tangential perturbation is chosen sufficiently large for the set of equilibria to be empty (see section 6.3.3).

The main result of this section can qualitatively be stated as follows: when submitted to a periodic forcing of sufficiently large amplitude, the behavior of the oscillator is such that there exist periodic solutions whatever the period of the excitation. Moreover:

i) for small periods, a single periodic solution exists, the period of which is that of the excitation and as the period increases the amplitude of the periodic solutions increases;

ii) for large periods, there also exists a single periodic solution, the period of which is that of the excitation but its amplitude does not depend on the period (in fact the amplitude is constant);

iii) between these two ranges, there exists an interval on the period axis in which at least one periodic solution coexists with another solution, which may exhibit chaos.

Some parts of the proof of this global result are essentially in progress at the time this book was published and the justifications rely on numerical experiments, but some other parts already have been based upon theoretical tools. The current state of these justifications is now presented.

6.5.1. *For large periods*

When the period of the forcing is large, the single periodic solution can be constructed from an initial data with a zero velocity. The orbit of this periodic orbit involves a cusp and is represented in Figure 6.13.

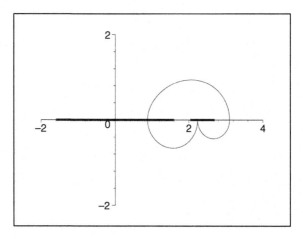

Figure 6.13. *The single periodic solution for T sufficiently large (the thick intervals on the horizontal axis represent the set of equilibria Σ_0 and Σ_ε)*

PROPOSITION 6.2.– Assume the amplitude of the forcing ε is greater than $\bar{\varepsilon}$, which means that the set of equilibria is empty. Then, when the period of the forcing is sufficiently large:

i) there exists a periodic solution having the period of the excitation,

ii) the orbit of the periodic solution does not depend on T.

PROOF.–

Proof of (i): There exists a single periodic solution of period $2T$.

Numerical observations guided the direct construction that is used in the proof of this first step. It is formally very close to the proof of the corresponding result obtained in Chapter 6 in the case of linear stiffness, except for the important fact that because of the nonlinearity, implicit equations are to be solved.

In the proof of proposition 6.1, an integration procedure for the equation of the dynamics was used, either for positive or for negative sliding, with or without the perturbation of amplitude ε. The same tools will be used to build a periodic solution.

Starting in positive sliding from the initial condition $(y_0, 0)$, where y_0 belongs to the set Σ_0, and assuming that the value of the perturbing tangential force remains equal to ε for a sufficiently long time, the velocity comes to zero again at time t_1 at a position y_1, which satisfies the equation:

$$(y_0^2 - y_1^2) - \frac{2\varepsilon(y_0 - y_1)}{h}$$

$$-2\beta(\sqrt{1 + y_0^2} - \sqrt{1 + y_1^2}) + 2\beta\mu(argsh(y_0) - argsh(y_1)) = 0.$$

Then, starting from the initial condition $(y_1, 0)$, for a negative sliding with the same value ε of the perturbation, the velocity comes to zero again at time t_2 and at position y_2. So, y_2 is the solution of the following equation:

$$(y_1^2 - y_2^2) - \frac{2\varepsilon(y_1 - y_2)}{h}$$

$$-2\beta(\sqrt{1 + y_1^2} - \sqrt{1 + y_2^2}) - 2\beta\mu(argsh(y_1) - argsh(y_2)) = 0.$$

Finally, starting from the initial condition $(y_2, 0)$, still in negative sliding, but where the perturbation has changed to zero, the velocity comes to zero again at time t_3 and at position y_3 given by the equation:

$$(y_2^2 - y_3^2) - 2\beta(\sqrt{1 + y_2^2} - \sqrt{1 + y_3^2}) - 2\beta\mu(argsh(y_2) - argsh(y_3)) = 0.$$

This construction is valid as long as $T \geq (t_1 + t_2)$ and y_2 belongs to the set Σ_ε and as long as y_3 belongs to the set Σ_0. A mapping ϕ from Σ_0 into Σ_0 such that $\phi(y_0) = y_3$ is thus defined. It is clear that a periodic solution exists if and only if there exists $\tilde{y}_0$ such that $\phi(\tilde{y}_0) = \tilde{y}_0$, i.e. $\tilde{y}_0$ is a fixed point of the mapping ϕ. As the mapping ϕ is continuous from the compact set Σ_0 into itself, there exists a fixed point to ϕ.

From a numerical point of view, this fixed point is obtained by usual fixed point iterations. The mapping ϕ is strictly decreasing in Σ_0 so this fixed point is unique. The numerical values chosen at the beginning of section 6.4 lead to $\tilde{y}_0 = 0.8736596$.

There exists a lower bound for the period of the forcing to lead to such a trajectory.

As observed above, this construction is possible only if $T \geq (t_1 + t_2)$. Let us denote this lower bound by T^{**}. The value of T^{**} can be explicitly calculated through equation [6.19]:

$$T^{**} = \int_{y_0}^{y_1} \frac{ds}{\sqrt{C_1 + \dfrac{2k}{m}\left[\left(\mu + \dfrac{F_t + \varepsilon - \mu F_n}{kh}\right)s - \dfrac{s^2}{2} - \mu\beta argsh(s) + \beta\sqrt{1 + s^2}\,\right]}}$$

$$+ \int_{y_1}^{y_2} \frac{-ds}{\sqrt{C_2 + \dfrac{2k}{m}\left[\left(\mu + \dfrac{F_t + \varepsilon + \mu F_n}{kh}\right)s - \dfrac{s^2}{2} + \mu\beta argsh(s) + \beta\sqrt{1 + s^2}\,\right]}}.$$

$$[6.23]$$

The numerical values chosen at the beginning of section 7.4 give $T^{**} = 6.874539$. The time t_3 is of course smaller than $t_1 + t_2$.

Proof of (ii): As soon as $T \geq T^{**}$, the shape of this periodic orbit in the phase plane does not depend on T.

Since y_2 and y_3 are, respectively, equilibrium states when the loading is perturbed by ε and when it is not, the mass remains at rest at these two points until the external force changes. The mass will remain at rest at point y_2 until the external force changes that is during $T - t_1 - t_2$. And the mass will stay at rest at point y_3 during $T - t_3$. So, the time intervals during which the mass stays at rest increase with T. The trajectories of the solutions depend on the value of the half-period T, see, for example, Figure 6.14 where the trajectories for $T = 8$ and $T = 16$ are represented. However, the orbits in the phase plane of these two solutions are identical and given by Figure 6.13.   □

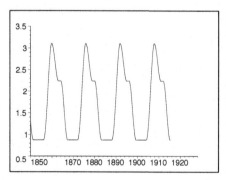 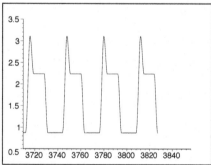

a) $T = 8$ b) $T = 16$

Figure 6.14. *The trajectory involves time intervals where the mass remains at rest*

6.5.2. *For small periods*

PROPOSITION 6.3.– Assume the forcing still has a sufficiently large amplitude for the set of equilibria to be empty, but now has a sufficiently high frequency. Then:

i) there exists a single periodic solution that has the period of the excitation;

ii) the amplitude of this periodic solution increases with T, as long as T does not overpass a critical value denoted by T^*.

A few trajectories illustrating proposition 6.3 are represented in Figure 6.15.

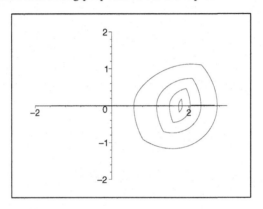

Figure 6.15. *Four successive small amplitude periodic solutions as T increases in the range $]0, T^*[$*

Toward the proof:

– Point (i) states that there exists a single periodic solution of period $2T$. The idea is to use the same direct construction as for proposition 6.2. The numerical experiments suggest that the solution should have a phase difference between the loading and the answer. Thus, the method used in the previous section in the case of large periods is now applied with an initial data at $t = 0$ equal to $(y(0) = y_0, \dot{y}(0) = v_0)$ where v_0 is strictly positive. The procedure is nevertheless essentially the same:

- let y_1 be the value of the displacement at time t_1 when the velocity of the positive sliding with perturbation ε reaches zero;

- let t_1 be smaller than T, then the motion continues in negative sliding with the perturbation ε up $(y_2(t_2), \dot{y}_2(t_2))$, where $t_2 = T$;

- let the perturbation ε be removed at time t_2 and y_3 be the value of the displacement at time t_3 when the velocity reaches zero again after a phase of negative sliding;

- finally, the motion continues in positive sliding without perturbation on the loading up to time $2T$. Let $(y_4, \dot{y}_4)$ be the displacement and velocity at that time $2T$.

Therefore, a periodic solution of period $2T$ exists if:

$$y_4(y_0, v_0) = y_0, \quad \dot{y}_4(y_0, v_0) = v_0. \qquad [6.24]$$

This procedure defines a mapping from $\mathbb{R}^2$ into $\mathbb{R}^2$, and a periodic solution exists when this mapping has a fixed point.

– The second point of proposition 6.3 states that the amplitude of the periodic solution increases from $T = 0$ monotonically with T. It was observed through numerical computations together with the existence of an upper bound of the range of the periods where this behavior is observed. Let T^* be this upper bound, which is observed to be strictly smaller than the lower bound T^{**} obtained in the previous section. The behavior of solutions for periods larger than T^* is described in the next section.

In the case of the numerical values adopted in this chapter, the half-period T^* is found to be around $T^* = 3.2$. Indications for the reason of the very weak accuracy of this result will be given in the next section.

6.5.3. *In the interval* $]T^*, T^{**}[$

This last section is based only on numerical experiments, but it seems that things really become more intricate in the range of periods $]T^*, T^{**}[$. The calculations are performed very rigorously as described in section 6.4.1, but the question of whether theoretical foundations could be obtained for what is observed is open for the moment. The main qualitative features that have been observed are the following:

i) There are apparently two simultaneous phenomena that appear at $T = T^*$. The first phenomenon is that the periodic solution of period equal to the period of the excitation computed in proposition 6.3 splits into two loops so its period becomes twice that of the excitation. The second phenomenon is that a loss of uniqueness of the periodic solution is observed, characterized by the occurrence of a solution of very large amplitude, consequently involving very large velocities. The corresponding global behavior is represented in Figure 6.16. It is very difficult to compute this value T^* precisely because it is difficult to determine exactly when the period becomes double (in fact only a lower bound for T^* can be given) and because the computational time needed to compute solutions of large amplitude increases drastically as T approaches T^* (so here only an upper bound for T^* can be given).

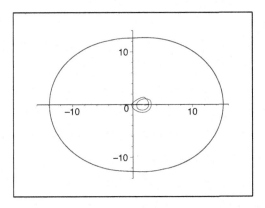

Figure 6.16. *Two periodic trajectories for $T \in \,]T^*, T^{**}[$
(in this figure $T = 3.45$)*

ii) As T increases, the response of the system changes in the following way:

- the amplitude of the very large amplitude solution decreases abruptly, as represented in Figure 6.17. Figure 6.17(b) shows that the amplitude tends to infinity as the period decreases to $2T^*$;

- on the other hand, the evolution of the double period solution seems to involve an accumulation of period doubling, which, in the case of smooth dynamical systems, would announce a transition toward chaos. This is qualitatively represented in Figures 6.18 and 6.19. Moreover, the computations reveal that this qualitative behavior appears, disappears and appears again several times in small sub-intervals of $]T^*, T^{**}[$.

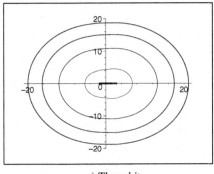

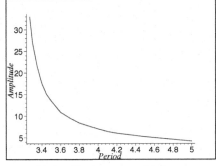

a) The orbits   b) Amplitude versus period

Figure 6.17. *Decreasing of the size of the orbit of the large amplitude
periodic solutions*

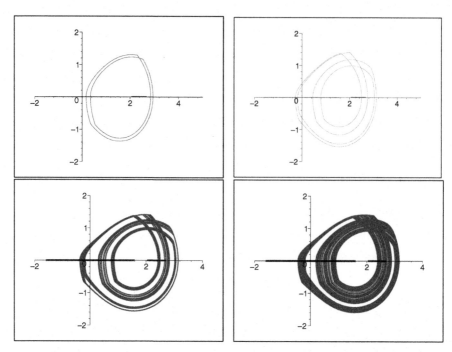

Figure 6.18. *From the periodic solution to the strange attractor, where successively* $T = 3.3$, $T = 3.5$, $T = 3.55$, $T = 3.59$

iii) For larger periods, the very complicated attractor seems to disappear, while a cusp appears on the large amplitude solution that tends to the single solution given in proposition 6.2 when T reaches T^{**}.

A part of the new qualitative features that have been exhibited in this chapter has been observed numerically. What occurs in the range (T^*, T^{**}) is far from having been fully investigated. This is the case for the second point of proposition 6.3 together with the whole section 7.5.3, which presents the transition to chaos and the shape of the attractors.

However, these results are of special importance in the context of dynamical systems. The transition to chaos must be observed in the light of the coupling between smooth and nonsmooth nonlinearities since no such transition was observed while investigating the dynamics when the restoring force was linear. The next chapter will discuss these questions further and state open problems together with some suggestions of possible future works.

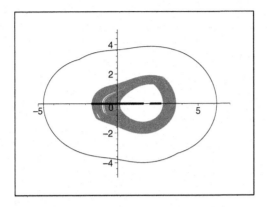

Figure 6.19. *The large amplitude periodic solution of period* $2T$
coexists with the chaotic attractor issued from the fundamental solution

7

Open Problems and Challenges

This chapter summarizes the results presented in this book, then suggests different types of future work that could be considered as complementary to the book. The first line of study consists of adding results from sections that have been dropped or removed from the book, either to prevent overloading, or, when no remaining difficulties persist, because those additional calculations are more or less identical to those already presented in the book. Such problems cannot therefore be considered as open. The second line of study really consists of open problems. Different parts of the book give rise to these open problems and some of them might be both long term and difficult ones.

7.1. Complementary calculations

7.1.1. What has been done?

A qualitative analysis of dynamical systems usually involves an exploration of the behavior of the solutions for any values of the frequency and amplitude of the forcing:

– the first part of this exploration consists of investigating the equilibrium states in the case of simple low dimensional systems:

- this investigation has been completely and explicitly performed when the restoring force is linear;

- in the case of a more general restoring force, either the investigation of the equilibrium states has been explicitly given, or it could be achieved by maintaining the strategy of separately studying the existence of states in contact and of states without contact. The additional nonlinearity that has been studied here arises from large deformations, but other nonlinearities, for example, due to nonlinear elasticity, would not lead to new difficulties.

– the second part of the exploration concerns the response under oscillating loading. A partition of the {$period,\ amplitude$} plane of the excitation, which is a fundamental step in the qualitative analysis of dynamical systems, has been obtained in the case of a linear restoring force. These qualitative results establish that in a range of the plane there exist only periodic solutions, non-periodic solutions, which converge toward the periodic ones, and no equilibrium states; while in another range there exist only equilibrium states and trajectories, which converge toward the equilibria, some of them in finite time, but no non-trivial periodic solutions. Periodic solutions have been found everywhere in the part of the {$period,\ amplitude$} plane where the amplitude is large enough for equilibrium not to exist, but non-periodic solutions that do not converge toward an equilibrium state or a periodic solution have not been found in the case of a linear restoring force, i.e. when the only nonlinearity is due to contact and friction.

In the case of a nonlinear restoring force specific qualitative properties appear, such as cascades of period doubling of the sliding periodic solutions, the fact that under particular values of the loading no trajectory can lose contact even when a large amplitude forcing is applied.

7.1.2. *Complementary calculations for higher dimensional problems*

A large part of the book is concerned with the behavior of a mechanical system consisting of a single mass moving in the plane. A two-mass system has been investigated only in the linear case, in order to establish conditions for a trajectory starting from one equilibrium to attain another equilibrium after a perturbation of the force. This has been done to support a stability conjecture. For systems of a much larger size, the general result that has been obtained and used in this book, namely that the set of equilibria in contact belongs to the intersection of the graph of the friction law with a manifold given by the equilibrium equation – an affine manifold in the linear case – still holds regardless of the size of the problem. But it has been observed that simply passing from one mass to two masses drastically increases the complexity of the investigation, so that an explicit exploration of the set of equilibria of a system with a large number of degrees of freedom is out of the question. Moreover, although the theoretical results of Chapter 3 establish well-posedness conditions and consequently justify the computation of all the possible dynamics for any finite dimensional system, explicit calculations have been presented here only in one case. As for numerical experiments, a large-size system has been presented in the case of slightly more than 600 degrees of freedom. This finite dimensional calculation is supposed to represent the dynamics of an elastic block impacted by a rigid projectile. A more specific investigation concerns granular media, systems with no stiffness matrix, but other concepts of equilibrium states and trajectories of very large-size systems with contact and friction are needed, such as those proposed by Jean *et al.* [JEA 09].

7.1.3. *Investigating behavior under general excitations*

In dynamics, the analysis has essentially been performed in the case of a piecewise constant loading, the so-called rectangular wave shape. Such a particular excitation was chosen with the intention of being didactic, in particular by carrying out explicit analytical calculations much further than could have been performed with a more general shape. However, it is clear that this special loading may lead to nonsmooth orbits that would not appear otherwise. Any smooth function, such as a sine function, requires numerical calculations and, although probably closer to the physics, would remove a part of the didactic intention. Numerical calculations have nevertheless been carried out in some cases, either with a sine function in the case of the simple one mass model in Chapter 6 but without performing a full investigation, or in the case of the elastic block impacted by a rigid projectile where the impact, which is a kinematic condition, was replaced for some experiments by a high-frequency sine function multiplied by a short length bell-shape function. A transient loading of this fashion was widely used by Jean while studying the shaking of a box filled with granular media [JEA 09].

7.1.4. *Investigation in the presence of impacts*

When the restoring force is linear, the boundary between a range which possesses sliding periodic solutions and another range where any initial data leads to a trajectory that loses contact is meticulously determined. In the latter range, all trajectories are no longer sliding but involve impacts and jumps. Are there still periodic solutions in this range? Numerical experiments suggested that the answer is yes in general. Moreover, no range containing no periodic solutions has been observed numerically. A general existence result, based on an implicit function argument, would provide an interesting result from the point of view of qualitative behavior of nonsmooth dynamics.

7.2. Toward more challenging problems

7.2.1. *Enlarging the theory of stability of discrete systems*

In Chapter 1, the need for a general stability theory to attain a complete understanding of the behavior of systems involving unilateral contact and Coulomb friction was expressed. Establishing such a stability theory is a challenging problem. The so-called classical stability theory was used to derive stability results by direct calculations of the trajectories. It was unfortunately not possible to refer to an energy argument, as given by the Lejeune-Dirichlet theorem (see, for example, [APP 04]) of which a more modern statement reads [ARN 74].

THEOREM 7.1.– Let a point q_0 be a strict local minimum of the potential energy U. Then, the equilibrium position $q = q_0$ is Lyapunov stable.

This impossibility is due to the fact that Coulomb friction removes the existence of a potential energy. In order to take Coulomb friction into account, the stability problem has been partially transformed by the introduction of a stability conjecture, which is based upon other arguments and is probably closer to physics. But, although this conjecture was backed up by several calculations, and although it is in agreement with many numerical computations and physical experiments, its proof in general seems out of reach, at least in the short term. This is due to the increase in the complexity with the size of the dynamical problem, although specific post-processing may be used to deduce stability properties in the sense of the conjecture for large-size systems.

The conjecture that has been given was formulated in the following way.

CONJECTURE 7.1.– Let a discrete system with any finite number of degrees of freedom be at equilibrium under unilateral contact and Coulomb friction conditions. Assume this equilibrium state is such that some reactions are strictly inside the Coulomb cone, while the other reactions are in imminent sliding. Thus, the trajectory produced by any sufficiently small perturbation of the data leads to a new equilibrium where the number of reactions strictly inside the cone is larger than before the perturbation.

COMMENTS.–

– A state is said to be in imminent sliding when the particle is motionless but the reaction lies on the border of the cone, so that an infinitely small perturbation may set the particle into motion.

– It is equivalent to say that an equilibrium state $(U = U^{eq}, \dot{U} = 0)$ is not perturbed by a small enough external force or to say that its reaction is strictly inside the cone. This is the foundation of the conjecture.

– This conjecture has been partially justified, in fact proved in the case of simple systems, and observed to be in agreement with a large-size numerical computation.

A simpler statement reads:

CONJECTURE 7.2.– Let a discrete mass–spring system with unilateral contact and Coulomb friction at equilibrium be perturbed by a sufficiently small constant force. Then, the trajectory leads in finite time to an equilibrium where all the reactions are strictly inside the Coulomb cone.

7.2.2. Improving the qualitative theory of dynamical systems

The qualitative behavior when smooth and nonsmooth nonlinearities are coupled is not as yet fully understood. Indeed:

– the contact and friction nonlinearities alone lead to a very unusual set of equilibrium states. For any classical dynamical system, the set of equilibrium states consists of a discrete set of points, but here, due to the fact that the right-hand sides of the equations of the dynamics involve mathematical objects which are not functions, the set of equilibrium states consists generically of an interval, possibly unbounded. This set can, nevertheless, be computed and is completely determined by the stiffness parameters, the friction coefficient and the components of the external force;

– adding contact and friction to a geometrical nonlinearity drastically change the set of equilibria. Indeed, the set of equilibria contains infinitely many points that completely fill one or several intervals, whereas in the case without friction, where the nonlinearity is due to large deformations alone, the set of equilibria consists of three separate points. In the case where the nonlinearity is due to contact and friction and the geometry is linear, the set of equilibria fills a single bounded or unbounded interval. Although very intricate, all these different situations have now been completely explored;

– the dynamics of the geometrically nonlinear oscillator are far from being completely understood. The new and fundamental question that arises from the investigation is the following: which characteristics of the nonlinearity lead to chaos?

In fact, insofar as the numerical calculations that have been performed are reliable, the complete investigation concerning the linear geometry, as given in Chapter 6, never shows a transition to chaos, even when nonlinearity due to impacts are added to the Coulomb friction nonlinearity (both being extremely strong nonlinearities). Conversely, if geometrical – that is, smooth – nonlinearity is added to friction, then chaos appears when the frequency of the excitation belongs to a given interval; but chaos was obtained with the same geometry without friction. The dynamical problem without unilateral nonlinearity looks very much like Duffing's equation (see [CAO 06]), which is obviously not the case when Coulomb friction and unilateral contact are introduced.

7.2.3. Refining the well-posedness criterion

It was recalled in Chapter 3 that existence of a trajectory holds as soon as the external force is given by an integrable function, but it was also stated that even when the external force is given by an infinitely differentiable function, the problem is ill-posed. Several counter examples were exhibited showing that the dynamical problem does not possess a unique solution, and it was recalled that uniqueness was recovered if the loading function is not only infinitely differentiable but is analytical. Does well-posedness occur when the loading is not an analytical function but is, nevertheless, strictly smoother than $\mathbb{C}^\infty$? The answer is yes and a provisional statement of the refined result can be written as [LUO 16]:

Let $F(t)$ be any infinitely differentiable function of time defined on an interval $[0, a)$ satisfying $F(0) = 0$ and $F^{(n)}(0) = 0$ for any n.

1) Assume F belongs to any quasi-analytic class, and assume that the initial data of problem [2.2] are both equal to zero, then the only solution to problem [2.2] is $U(t) \equiv 0$.

2) Assume F does not belong to a quasi-analytic class. Then, there exists a non-zero solution to problem [2.2].

The complete proof of this result is not as yet available.

7.2.4. And what about the continuous case...

The continuous case has also been evoked in the introductory chapter of this book. It is probably the problem for which long-term investment is most required. In the case of a discrete problem, the functional framework in time involves measures, with values in $\mathbb{R}$ for a single degree of freedom or with values in $\mathbb{R}^n$ for a more general system. But this framework has no meaning when the number of degrees of freedom goes to infinity, i.e. for a continuous body. In the case of linear elasticity in a domain Ω of smooth boundary $\partial\Omega \equiv \Gamma$, it is known that the natural framework of the equilibrium problem when Ω is submitted to square integrable forces and bilateral boundary conditions is the Sobolev space $\mathbb{H}^1(\Omega)$. But even in the static case, when the body is submitted to unilateral contact and Coulomb friction, the three following difficulties arise, as noticed in [NAE 96]:

– the normal component of the Cauchy stress tensor at the boundary $\sigma_N(U)$, which is needed to express unilateral conditions similar to those used throughout this book, is not defined for all displacement fields u in $\mathbb{H}^1(\Omega)$;

– let Γ_C be the part of the boundary Γ which is in contact with the obstacle. Then, the restriction to Γ_C of $\sigma_N(U)$ does not belong in general to $\mathbb{H}^{-1/2}(\Gamma]$, which is the natural framework in the case of bilateral linear elasticity, but to $(\mathbb{H}_{00}^{1/2})'$, where $\mathbb{H}_{00}^{1/2}$ is a subset of $\mathbb{H}^{1/2}$, the elements of which decrease sufficiently fast on the border of Γ_C;

– let μ be the friction coefficient defined on Γ_C. Then, the quantity $\mu\sigma_N(U)$ that necessarily appears in the variational formulation of Coulomb's law exists only if μ is sufficiently smooth.

These three difficult technical points of functional analysis have now been solved so that the static or the quasi-static problem has a correct variational formulation. But the existence and the uniqueness of a solution to these problems have not been obtained in the general case at this point in time, although a few partial results can be found. However, the dynamical problem with unilateral contact and friction remains an entirely open problem.

Bibliography

[ALA 86] ALART P., CURNIER A., "Contact discret avec frottement: unicité de la solution, convergence de l'algorithme", Publications du Laboratoire de Mécanique Appliquée, Ecole Polytechnique Fédérale de Lausanne, Lausanne, 1986.

[APP 04] APPELL P., *Traité Mécanique Rationnelle*, vol. 2, Gauthier-Villars, Paris, 1904.

[ARN 74] ARNOLD V., *Méthodes mathématiques de la mécanique classique*, Edition MIR, Moscow, 1974.

[BAL 05] BALLARD P., BASSEVILLE S., "Existence and uniqueness for dynamical unilateral contact with Coulomb friction: a model problem", *Mathematical Modeling and Numerical Analysis*, vol. 39, no. 1, pp. 57–77, 2005.

[BAL 00] BALLARD P., "The dynamics of discrete mechanical systems with perfect unilateral constraints", *Archive for Rational Mechanics and Analysis*, vol. 154, pp. 199–274, 2000.

[BAL 06] BALLARD P., LÉGER A., PRATT E., "Stability of discrete systems involving shocks and friction", in WRIGGERS P., NACKENHORST U. (eds), *Analysis and Simulation of Contact Problems*, Lecture Notes in Applied and Computational Mechanics, vol. 27, Springer, Berlin, Heidelberg, pp. 343–350, 2006.

[BAS 03] BASSEVILLE S., LÉGER A., PRATT E., "Investigation of the equilibrium states and their stability for a simple model with unilateral contact and Coulomb friction", *Archive of Applied Mechanics*, vol. 73, pp. 409–420, 2003.

[BAS 06] BASSEVILLE S., LÉGER A., "Stability of equilibrium states in a simple system with unilateral contact and Coulomb friction", *Archive of Applied Mechanics*, vol. 76, no. 7/8, pp. 403–428, 2006.

[BAS 00] BASTIEN J., LAMARQUE C.-H., SCHATZMAN M., "Study of some rheological models with a finite number of degrees of freedom", *European Journal of Mechanics A/Solids*, vol. 19, pp. 277–307, 2000.

[BRE 73] BREZIS H., *Operateurs maximaux monotones et semi-groupes de contraction dans les espaces de Hilbert*, North Holland, 1973.

[CAO 06] CAO Q.J., WIERCIGROCH M., PAVLOVSKAIA E., *et al.*, "An archetypal oscillator for smooth and discontinuous dynamics", *Physical Review*, vol. E 74, p. 046218, 2006.

[CAO 08a] CAO Q.J., WIERCIGROCH M., PAVLOVSKAIA E., *et al.*, "Piecewise linear appoach to an archetipal oscillator for smooth and discontinuous dynamics", *Philosophical Transactions of the Royal Society*, A, vol. 366, no. 1865, pp. 635–652, 2008.

[CAO 08b] CAO Q.J., WIERCIGROCH M., PAVLOVSKAIA E., *et al.*, "The limit case response of the archetypal oscillator for smooth and discontinuous dynamics", *International Journal of Nonlinear Mechanics,* vol. 43, pp. 462–473, 2008.

[CAO 15] CAO Q.J., LÉGER A., WIERCIGROGH M., *A Smooth and Discontinuous Dynamical System: Theory and Applications*, Springer, Beijing, 2015.

[CHA 14] CHARLES A., BALLARD P., "Existence and uniqueness of solution to dynamical unilateral contact problems with Coulomb friction: the case of a collection of points", *Mathematical Modelling and Numerical Analysis*, vol. 48, no. 1, pp. 1–25, 2014.

[CHO 98] CHO H., BARBER J.R., "Dynamics behavior and stability of simple frictional systems", *Mathematical Computer Modeling*, vol. 28, pp. 37–53, 1998.

[CIM 96] CIMETIÈRE A., LÉGER A., "Some problems about elastic-plastic post-buckling", *International Journal of Solids Structures*, vol. 32, no. 10, pp. 1519–1533, 1996.

[DUB 90] DUBOIS F., LMGC 90, available at http://www.lmgc.univ-montp2.fr, 1990.

[ELK 96] ELKOULANI A., LÉGER A., "Solutions bifurquées du problème en vitesses initiales pour une poutre élastoplastique", *Comptes Rendus Académie Sciences*, vol. 322, no. I, pp. 1007–1013, 1996.

[ELK 97] ELKOULANI A., LÉGER A., "Comportement post-critique des poutres élastoplastiques: existence et régularité des branches bifurquées", *Comptes Rendus Académie Sciences*, vol. 324, no. I, pp. 1307–1313, 1997.

[FIC 63] FICHERA G., "Sul problema elastostatico di Signorini con ambigue condizioni al contorno", *Atti della Accademia Nazionale dei Lincei Classe di Scienze Fisiche, Matematiche e Naturali, Serie VIII*, vol. 34, no. 2, pp. 138–142, 1963.

[GEY 00] GEYMONAT G., LÉGER A., "Nonlinear spherical caps and associated plate and membrane problems", *Journal of Elasticity*, vol. 57, pp. 171–200, 2000.

[JEA 85] JEAN M., THOULOUZE-PRATT E., "A system of rigid bodies with dry friction", *International Journal of Engineering Science*, vol. 23, no. 5, pp. 497–513, 1985.

[JEA 99] JEAN M., "The nonsmooth contact dynamics method", *Computer Methods in Applied Mechanics and Engineering*, vol. 177, pp. 235–257, 1999.

[JEA 09] JEAN M., CAMBOU B., RADJAI F., *Micromechanics of Granular Media*, ISTE, London and Wiley, Hoboken, 2009.

[KIM 89] KIM J.U., "A boundary thin obstacle for a wave equation", *Communications in Partial Differential Equations*, vol. 14, pp. 1011–1026, 1989.

[KIT 05] KITTEL Ch., *Introduction to Solid State Physics*, 8th ed., Wiley, 2005.

[KLA 90] KLARBRING A., "Examples of non-uniqueness and non-existence of solutions to quasistatic contact problems with friction", *ING. Archives*, vol. 60, pp. 529–541, 1990.

[LEB 84] LEBEAU G., SCHATZMAN M., "A wave problem in a half space with a unilateral constraint at the boundary", *Journal of Differential Equations*, vol. 53, pp. 309–361, 1984.

[LUO 16] LUO F.S., LÉGER A., "On the transition to well-posedness in the one dimensional impact problem", forthcoming, 2016.

[LÉG 11] LÉGER A., PRATT E., "Qualitative analysis of a forced nonsmooth oscillator with contact and friction", *Annals of Solid and Structural Mechanics*, vol. 2, nos. 1–17, 2011.

[LÉG 12] LÉGER A., PRATT E., CAO Q.J., "A fully nonlinear oscillator with contact and friction", *Nonlinear Dynamics*, vol. 70, pp. 511–522, 2012.

[LEI 03] LEINE R.I., GLOCKER Ch., VAN CAMPEN D.H., "Nonlinear dynamics and modeling of some wooden toys with impact and friction", *Journal of Vibration Control*, vol. 9, pp. 25–78, 2003.

[MAR 94] MARTINS J.A.C., MONTEIRO MARQUES M.D.P., GASTALDI F., "On an example of non-existence of solution to a quasistatic frictional contact problem", *European Journal of Mechanics A/Solids*, vol. 13, no. 1, pp. 113–133, 1994.

[MON 94] MONTEIRO MARQUES M.D.P., "An existence, uniqueness and regularity study of the dynamics of systems with one-dimensional friction", *European Journal of Mechanics A/Solids*, vol. 13, no. 2, pp. 277–306, 1994.

[MON 93] MONTEIRO MARQUES M.D.P., "Differential inclusions in nonsmooth mechanical problems", *Progress in Nonlinear Differential Equations and Their Applications*, Birkhäuser, Basel, 1993.

[MOR 88] MOREAU J.J., "Unilateral contact and dry friction in finite freedom dynamics", in MOREAU J.J., PANAGIOTOPOULOS P.D. (eds), *Nonsmooth Mechanics and Applications*, CISM Courses and Lectures 302, pp. 1–81, Springer-Verlag, Vienna/New York, 1988.

[NAE 96] NAEJUS C., CIMETIÈRE A., "About the variational formulation of the Signorini problem with Coulomb friction", *Comptes Rendus Académie Sciences*, vol. 323, no. I, pp. 307–323, 1996.

[PER 85] PERCIVALE D., "Uniqueness in the elastic bounce problem", *Journal of Differential Equations*, vol. 56, pp. 206–215, 1985.

[PIN 01] PINTO DA COSTA A.M.F., Instabilidades e bifurcacoes em sistemas de comportamento no-suave, PhD Thesis, Instituto Superior Técnico, Technical University of Lisbon, 2001.

[PIN 04] PINTO DA COSTA A.M.F., MARTINS J.A.C., FIGUEREDO I.N., *et al.*, "The directional instability problem in systems with frictional contact", *Computer Methods in Applied Mechanics and Engineering*, vol. 193, pp. 357–384, 2004.

[PRA 10] PRATT E., LÉGER A., JEAN M., "About a stability conjecture concerning unilateral contact with friction", *Nonlinear Dynamics*, vol. 59, pp. 73–94, 2010.

[PRA 06] PRATT E., LÉGER A., "Exploring the dynamics of a simple system involving Coulomb friction", *3rd European Conference Computational Mechanics Solids, Structures and Coupled Problems in Engineering*, Lisbon, Portugal, 5-9 June 2006.

[PRA 12] PRATT E. LÉGER A., CAO Q.J., "A fully nonlinear oscillator with contact and friction", *Nonlinear Dynamics*, vol. 70, pp. 511–522, 2012.

[PRA 08] PRATT E., LÉGER A., JEAN M., "Critical oscillations of mass-spring systems due to nonsmooth friction", *Archive Appl. Mech.*, vol. 78, pp. 89–104, 2008.

[PRA 13] PRATT E., LÉGER A., ZHANG X., "Study of a transition in the qualitative behavior of a simple oscillator with Coulomb friction", *Nonlinear Dynamics*, vol. 74, pp. 517–531, 2013.

[SCH 78] SCHATZMAN M., "A class of differential equations of second order in time", *Nonlinear Analysis*, vol. 2, pp. 355–373, 1978.

[SCH 98] SCHATZMAN M., "Uniqueness and continuous dependence on data for one dimensional impact problems", *Mathematical Computational Modeling*, vol. 28, pp. 1–18, 1998.

[SIG 59] SIGNORINI A., "Questioni di Elasticità non linearizzata e semilinearizzata", *Rendiconti di Matematica e delle sue applicazioni*, vol. 5, no. 18 pp. 95–139, 1959.

[TIA 10] TIAN R.L., CAO Q., LI Z.X., "Hopf bifurcation for the recently proposed SD oscillator", *Chinese Physics Letters*, Chinese Physical Society and IOP Publishing, 2010.

Index

Printed in the United States
By Bookmasters